SpringerBriefs in Food, Health, and Nutrition

Springer Briefs in Food, Health, and Nutrition present concise summaries of cutting edge research and practical applications across a wide range of topics related to the field of food science.

Editor-in-Chief
Richard W. Hartel
University of Wisconsin—Madison, USA

Associate Editors
J. Peter Clark, *Consultant to the Process Industries, USA*
David Rodriguez-Lazaro, *ITACyL, Spain*
David Topping, *CSIRO, Australia*

For further volumes:
http://www.springer.com/series/10203

Guillermo López-Campos
Joaquín V. Martínez-Suárez
Mónica Aguado-Urda · Victoria López-Alonso

Microarray Detection and Characterization of Bacterial Foodborne Pathogens

 Springer

Guillermo López-Campos
Department of Bioinformatics
and Public Health
Spanish National Institute
of Health 'Carlos III'
Majadahonda, Madrid, Spain

Mónica Aguado-Urda
Department of Animal Health
Complutense University
Madrid, Spain

Joaquín V. Martínez-Suárez
Department of Food Technology
Spanish National Institute for Agricultural
and Food Research and Technology (INIA)
Madrid, Spain

Victoria López-Alonso
Department of Bioinformatics
and Public Health
Spanish National Institute
of Health 'Carlos III'
Majadahonda, Madrid, Spain

ISBN 978-1-4614-3249-4   e-ISBN 978-1-4614-3250-0
DOI 10.1007/978-1-4614-3250-0
Springer New York Dordrecht Heidelberg London

Library of Congress Control Number: 2012932853

Printed on acid-free paper

Springer is part of Springer Science+Business Media (www.springer.com)

Acknowledgements

The financial support of the Spanish Ministry of Science and Innovation (Research Project RTA2008-00080-C02 and RETIC COMBIOMED) is gratefully acknowledged.

Contents

Chapter 1
Introduction to Foodborne Diseases

Abstract Death and disease caused by food lacking safety guarantees represent a continuing threat to worldwide public health and socioeconomic development. Recent large outbreaks of foodborne infections have alerted us to the possible increase in the incidence of foodborne diseases. The epidemiology of these diseases has changed in recent decades, not only due to the emergence of new pathogens but also due to changes that have occurred in the food supply, including changes in processing and consumer preference and a global marketplace, and by the increase in populations with greater susceptibility to these diseases. Epidemiological surveillance, research into new problems, and collaboration among the different professional groups to put measures for controlling foodborne diseases into practice are the main strategies that are becoming increasingly necessary. The possibilities offered by modern tools and molecular techniques have greatly expanded the options available for monitoring undesirable microorganisms in the food supply chain and for quickly implementing appropriate measures as a result of the rapid detection and characterization of pathogens.

Keywords Food safety • Foodborne infections • Outbreaks of foodborne illness • Emerging pathogens • Foodborne disease surveillance • Foodborne disease control

1.1 Foodborne Diseases Remain a Significant Cause of Morbidity and Mortality

Death and disease caused by food lacking safety guarantees represent a continuing threat to worldwide public health and socioeconomic development. Humans acquire infections from a large variety of sources and by various transmission routes. Most foodborne diseases are mild and are associated with acute gastrointestinal symptoms, such as diarrhea and vomiting. In some cases, however, foodborne diseases

G. López-Campos et al., *Microarray Detection and Characterization of Bacterial Foodborne Pathogens*, SpringerBriefs in Food, Health, and Nutrition, DOI 10.1007/978-1-4614-3250-0_1, © Guillermo Lopez-Campos, Joaquin V. Martinez-Suarez, Mónica Aguado-Urda, Victoria Lopez Alonso 2012

can be more severe and may even be life-threatening, especially in children in developing countries. These diseases may also cause chronic sequelae and disabilities (Blackburn and McClure 2009).

The true extent of the cost of all types of foodborne diseases associated with microorganisms (bacteria, viruses, and parasites) is not precisely known, but it is thought to be enormous (Kuchenmüller et al. 2009; Newell et al. 2010). Despite the enormous effort in recent decades to prevent and control foodborne diseases, people in both industrialized and developing countries continue to suffer from them in large numbers. In the United States alone, the cost incurred by foodborne diseases in 2007 was estimated to be between $357 billion and $1.4 trillion (Roberts 2007). Developing countries tend to suffer from the largest share of the burden of foodborne diseases. The World Health Organization (WHO 2011) estimates that foodborne and waterborne diarrheal diseases kill about 2.2 million people annually, 1.9 million of them children. In addition to reducing the financial costs, public health spending will reduce morbidity and mortality associated with these diseases. It has been estimated that increasing the supply of rural health care by 65% in developing countries could save 1.2 million lives annually (Green et al. 2009).

1.2 Current Trends in Foodborne Diseases

Data on current trends in foodborne diseases are limited to a few industrialized countries and a surprisingly small number of pathogens (Blackburn and McClure 2009; Ammon and Makela 2010; Newell et al. 2010). Despite having better information systems than other regions, statistics on foodborne diseases in the US and some European countries are overly dominated by cases of salmonellosis and campylobacteriosis. In contrast, the statistics in other regions depend almost exclusively on information about disease outbreaks, with the result being that ultimately other microorganisms are identified as the main causes of these diseases.

Epidemiological surveillance and control of foodborne viral pathogens are usually scarce. In Western countries, noroviruses and hepatitis A virus are the main human foodborne pathogens in terms of the number of outbreaks and people affected. Most foodborne viral infections originate from infected people who handle food that is not subsequently heated or otherwise treated (Koopmans and Duizer 2004).

There are many known foodborne pathogenic parasites, such as *Ascaris*, *Cryptosporidium,* and *Trichinella*, but their presence in food, livestock, and the environment is often not investigated. This results in a lack of information about the epidemiology of parasite contamination throughout the food supply chain (Newell et al. 2010). According to the Centers for Disease Control and Prevention (CDC 2011), *Cryptosporidium* is reported to be one of the leading causes of laboratory-confirmed foodborne infections in the United States.

The main foodborne bacterial pathogens are *Salmonella*, *Campylobacter*, *Yersinia*, Shiga-toxin- (Stx) producing *Escherichia coli* (STEC) and *Listeria monocytogenes*. Antibiotic resistance is also a significant issue with some foodborne pathogenic bacteria (*Salmonella*, *Campylobacter*, *Shigella*, *Vibrio*, *Staphylococcus aureus*, *E. coli*, and *Enterococcus*). In 2009, campylobacteriosis, salmonellosis, and yersiniosis were the most frequent foodborne zoonotic infections in humans in the EU (Lahuerta et al. 2011). Infections due to *Salmonella* were the most common in the US and were associated with the highest number of hospitalizations and deaths among all foodborne diseases. *Campylobacter*, *Shigella*, and STEC O157:H7 are also leading causes of bacterial foodborne infections in the United States (CDC 2011).

According to the European Food Safety Authority (EFSA 2011), in the EU most of the 5,550 confirmed outbreaks of foodborne diseases were caused by *Salmonella*, viruses, and bacterial toxins. The foods most frequently involved were eggs, buffet-style meals, and pork. In an analysis of foodborne disease outbreak data to estimate the proportion of human cases of specific enteric diseases attributable to a specific food item, Greig and Ravel (2009) found that a few etiologies were very specifically associated with some foods. *Salmonella* Enteritidis outbreaks occurred relatively often in the EU, with eggs from laying hens the most common source (Pires et al. 2011). Detailed analysis of different regions highlighted some special features: *Campylobacter*-associated outbreaks were mainly related to poultry products in the EU and to dairy products in the US (Greig and Ravel 2009). STEC in cattle and beef and *Yersinia* in pigs and pork also represent significant associations between zoonotic infections and food in the EU (EFSA 2011). Also among the main concerns and challenges for food safety and hygiene are *L. monocytogenes* in ready-to-eat (RTE) processed food, and viral pathogens at food establishments (Sofos and Geornaras 2010).

Recent large outbreaks of foodborne infections have alerted us to the possible increase in recent years in the incidence of foodborne diseases. However, 2010 statistics from the US on the incidence of six significant foodborne pathogens show no significant changes when compared to the period 2006–2008. Only the infection rates for *Shigella* and STEC O157:H7 show significant declines, while those of *Vibrio* show significant increases (CDC 2011). In some cases, increased awareness about food safety, changes in legislation and educational measures, as well as changes in food production practices (e.g., effective control measures for animal reservoirs) have produced declines in the incidence of certain foodborne diseases in some regions (Blackburn and McClure 2009). For example, in 2009 the EU confirmed the declining trend in the number of salmonellosis cases in humans, with statistically significant figures (EFSA 2011). Campylobacteriosis represents the most frequent zoonosis in the EU (EFSA 2011), and *Campylobacter* spp. represents the main cause of sporadic bacterial gastroenteritis worldwide. However, in some countries the number of cases of campylobacteriosis seems to have stabilized or declined, and it has been suggested that this decline may be due to the implementation of the Hazard Analysis and Critical Control Point (HACCP) system in poultry industries (CDC 2011). In the case of listeriosis, a decline in the number of cases was also observed toward the end of the twentieth century. This has been linked to efforts carried out by food processing companies and food regulatory agencies to control *L. monocytogenes* in high-risk foods (Swaminathan and Gerner-Smidt 2007). However, there is currently growing concern about the increase in listeriosis cases in the elderly (those over 60 years of age), which likely explains the 19.1% increase in human listeriosis cases in the EU in from 2008 to 2009 (EFSA 2011).

Therefore, despite greater prevention efforts, foodborne diseases continue to be an ongoing problem in Europe and the US. However, the increase or decrease in its incidence appears to depend on the pathogen and the region being studied (Nyachuba 2010). The decline in foodborne diseases caused by particular pathogens can be related to the introduction of and adherence to effective control measures to minimize the risk of infection.

1.3   Emerging Foodborne Diseases

The microbiological safety of food is a dynamic issue subject to the powerful influence of numerous factors along the entire food supply chain from farm to table. The pathogen populations most relevant to food safety are not static, and food is an excellent vehicle through which many pathogens reach suitable places for colonizing new hosts (Newell et al. 2010). Consequently, the epidemiology of foodborne diseases has changed in recent decades, not only due to the emergence of new pathogens but also due to changes that have occurred in the food supply and by the increase in populations with greater susceptibility to these diseases (Altekruse and Swerdlow 1996).

The spectrum of foodborne infections has changed as some traditional pathogens have been controlled or eliminated while other previously unknown pathogens, many of them zoonotic, have emerged as human pathogens (Newell et al. 2010). Emerging diseases are often described as those whose prevalence have increased in recent decades or likely will increase in the near future (Altekruse and Swerdlow 1996). Therefore, an emerging pathogen does not necessarily have to have emerged recently. Even well-recognized foodborne pathogens, such as *Salmonella* spp. and enteropathogenic strains of *E. coli,* may evolve to take advantage of new situations. They may, for example, contaminate foods that were not previously considered at risk, such as fresh vegetables, and may cause new public health challenges, such as resistance to antibiotics. Both STEC O157:H7 (Inset 1.1) and antibiotic-resistant strains of *Salmonella* are found in cattle and are examples of pathogens that have evolved relatively recently (Blackburn and McClure 2009).

New pathogens may emerge due to the uptake of mobile virulence factors found in large regions of DNA known as *pathogenicity islands* (PAIs). These islands may be shared among various pathogens, contributing to their evolution (Ahmed et al. 2008). Genetic promiscuity, which allows these genes to be acquired, is related to the presence of a series of mobile genetic elements, such as plasmids, transposons, conjugative transposons, and bacteriophages (Fig. 1.1). Both DNA acquisition and genome reduction have important roles in genome evolution. *E. coli* O157:H7, for example, probably evolved from an enteropathogenic ancestor of serotype O55:H7 through horizontal gene transfer and recombination (Feng et al. 1998). In other cases, pathogens may emerge due to ecological or technological changes that bring

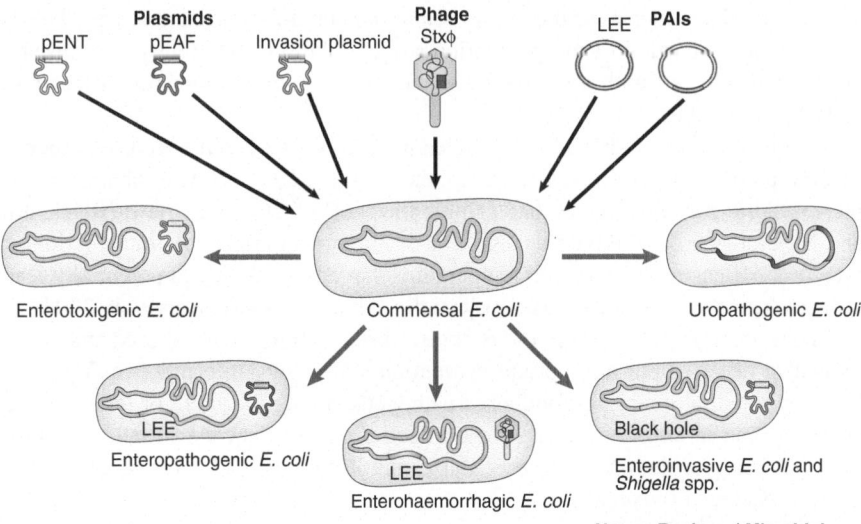

Nature Reviews | Microbiology

Fig. 1.1 Contribution of the horizontal acquisition of mobile genetic elements to the evolution of *Escherichia coli* pathotypes. *LEE* locus of enterocyte effacement, *PAI* pathogenicity island, *pEAF* enteropathogenic *E. coli* adhesion-factor plasmid, *pENT* enterotoxin-encoding plasmids, *Stx* Shiga-toxin-encoding bacteriophage (Reprinted by permission from Macmillan Publishers Ltd., Ahmed et al. (2008))

a potential pathogen into contact with the food supply chain. Relatively recent examples of pathogens associated with food as vehicles for disease include *L. monocytogenes*, *Cryptosporidium parvum*, and *Cyclospora cayetanensis*. Many of these foodborne pathogens are *zoonotic*: They have an animal reservoir without necessarily causing disease in that animal (Blackburn and McClure 2009).

Along with the emergence (or recognition) of new pathogens and resistance to antibiotics, several other trends are currently considered to be highly important in the evolution of microbiological food safety. These include the global pandemics of some foodborne pathogens, the identification of pathogens that are especially opportunistic and only affect human subpopulations at high risk, and the increasingly frequent identification of large and widely distributed disease outbreaks (Newell et al. 2010).

These emerging situations are often due to changes in some aspect of the social environment. The global economy, for example, facilitates the rapid transport of perishable food, increasing the chances of exposure to pathogens from other parts of the world (Käferstein et al. 1997). Other aspects related to food itself that change the patterns of foodborne diseases are the type of food that the population eats (consumer trends), the sources of these foods, and the possible decline in public awareness of safe food preparation practices (food security) (Blackburn and McClure 2009).

Demographic-associated factors, such as the aging population, obesity, the AIDS epidemic, increased life expectancy of the chronically ill thanks to medical technology, and so forth, have increased the public health impact of foodborne diseases. This is because the proportion of the population susceptible to developing severe disease after being infected with a foodborne pathogen continues to increase (Blackburn and McClure 2009).

Various factors have been hypothesized to promote the simultaneous emergence of new pathogens in different parts of the world. Many of these changes have increased the possible impact that a single focus of infection may have (Blackburn and McClure 2009). Climate change, for example, can affect the transmission of infectious diseases, as increased temperatures have been shown to potentially lead to increases in foodborne diseases, especially salmonellosis (Kovats et al. 2004).

The changing epidemiology of foodborne diseases must be monitored and investigated in order to put appropriate prevention technology into practice. We must look for potential emerging foodborne pathogens among the silent or less frequent zoonoses and among the more severe infections that affect the immunocompromised human population. As Robert Tauxe rightly said in 2002, "We should expect the unexpected" (Tauxe 2002).

1.4   Surveillance and Control of Foodborne Diseases

Most reports on zoonotic microorganisms and outbreaks of foodborne infections highlight the importance of having appropriate surveillance systems to monitor animals, food, and humans in order to constantly track the changing trends of well-known diseases and detect emerging pathogens (Lahuerta et al. 2011). It is also necessary to understand the complex interactions that these pathogens have with their environment during transmission throughout the food supply chain so as to develop effective prevention and control strategies (Newell et al. 2010).

Surveillance data are used for planning, implementing, and evaluating public health policies, as well as the identification of new hazards. The use of surveillance data is important for the education and training of individuals participating in the manufacture, handling, and preparation of food, including consumers. Therefore, there is a strong need to improve surveillance systems for foodborne disease. It is essential that surveillance systems share information across borders and promptly report foodborne disease outbreaks using modern telecommunication tools where possible (Blackburn and McClure 2009). The increasing possibilities offered by molecular techniques have greatly expanded the options available for detecting undesirable microorganisms in the food supply chain and for quickly implementing appropriate measures. The training of teams with experts from every field is crucial for achieving these objectives.

Food may be contaminated at any link of the food supply chain, from the farm to the table, including in the kitchen of the consumer. Therefore, measures to reduce and control the risk of contracting foodborne disease must be undertaken at each step in food preparation. Additionally, the complexity of the global food market shows us that the control of foodborne diseases must be a shared task, requiring action at all levels, from the individual to the international community. There are numerous ways of preventing most foodborne diseases, and the best approach is the simultaneous utilization of several measures. An example would be applying measures to eliminate microorganisms during the processing of food while at the same time reducing the likelihood of the microorganism's being present in the food at its source. It is also important for long-term prevention to understand how pathogens persist in animal reservoirs (for example, in herds) (Tauxe 2002). Preventive measures such as good manufacturing practices, supplemented by the HACCP system, have been introduced as a means of ensuring the production of safe food. However, their use does not necessarily provide quantitative information on the risks associated with the consumption of a particular food product. To obtain such information, elements of quantitative microbiological risk assessment (QMRA) need to be used. The four cornerstones of QMRA are hazard identification, exposure assessment, hazard characterization, and risk characterization. These steps represent a systematic process for identifying adverse consequences and their associated probabilities arising from consumption of foods that may be contaminated with microbial pathogens and/or microbial toxins (Lammerding and Fazil 2000; Rotariu et al. 2011).

Inset 1.1 Enterohemorrhagic Strains of *Escherichia coli* as a Paradigm of Emerging Pathogens

---

Escherichia coli is one of the commensal species in the human gut; however, several pathogenic *E. coli* strains have emerged that cause disease in humans. These strains are classified based on the disease they cause and their unique virulence factors (Feng and Weagant 2011). The different pathotypes include Enterotoxigenic *E. coli* (ETEC), Enteropathogenic *E. coli* (EPEC), Enterohemorrhagic *E. coli* (EHEC), Enteroinvasive *E. coli* (EIEC) (including *Shigella* spp.), extraintestinal pathogenic *E. coli* (such as Uropathogenic *E. coli*, UPEC), and others (Fig. 1.1). The first four groups have been implicated in food- or waterborne illness (Feng and Weagant 2011).

EHEC are zoonotic pathogens associated with both major outbreaks and sporadic cases of diarrhea and hemorrhagic colitis (HC) or bloody diarrhea. If not treated, HC can progress to hemolytic uremic syndrome (HUS), which can result in protracted illness or even death (Karmali et al. 2010). EHEC is noteworthy because a small but significant number of infected people develop HUS, which is the most frequent cause of acute renal failure in children in the Americas and Europe (Pennington 2010).

EHEC is a subgroup of STEC, also known as VTEC [verocytotoxin (VT)-producing *E. coli*]. There are many serotypes of STEC, but only those that have been clinically associated with HC are designated as EHEC. Of these, O157:H7 is the prototypic EHEC most often implicated in illness worldwide

(continued)

Inset 1.1 (continued)

(Feng and Weagant 2011; Karmali et al. 2010). *E. coli* O157 can be transmitted to humans by three primary (foodborne, environmental, waterborne) and one secondary (person-to-person) transmission pathways. According to Rotariu et al. (2011), QMRAs indicated that waterborne transmission was the least significant, but it was unclear whether food or the environment was the main source of infection.

EHEC serotypes other than O157:H7 possess the same main virulence factors as O157:H7 and are all frequently found in healthy cattle and other ruminants, causing little or no discernible disease in their animal reservoirs (Ferens and Hovde 2011). It is not clear whether all *E. coli* O157 isolates of animal origin are equally harmful to humans. In contrast to O157:H7, the pathogenic potential of atypical *E. coli* O157 isolates and several non-O157 serotypes often is ignored (Lefebvre et al. 2008).

It was predicted years ago that STEC strains other than the O157:H7 serotype would emerge as significant foodborne pathogens. Since then, these microorganisms have been linked to numerous outbreaks and sporadic cases of disease around the world (Blackburn and McClure 2009). The incidence of these serotypes continues to grow, which means they can be considered emerging pathogens (Mathusa et al. 2010). A recent example is the large outbreak of foodborne infections caused by EHEC O104:H4, which was mainly centered in Germany (Frank et al. 2011) and affected people in numerous European countries and the US, lasting throughout May and June of 2011 (Robert Koch Institute 2011). The total number of confirmed cases of STEC in Europe was 941 according to the European Center for Disease Control and Prevention (ECDC 2011). This figure includes 264 cases of HUS and a total of 46 deaths, marking one of the largest outbreaks of HUS ever described worldwide.

The pathogenesis of EHEC infection is multifactorial and involves several levels of interaction between the bacterium and the host. EHEC strains carry a set of virulence genes that encode factors for attachment to host cells, elaboration of effector molecules, and production of different toxins (Melton-Celsa et al. 2012). These genes are found in a large PAI called the locus of enterocyte effacement (LEE) located on the chromosome, in lambdoid phages, and in a large, approximately 60-MDa, virulence-associated plasmid, referred to as *EHEC plasmid* (Fig. 1.1). Genomic fluidity among these elements has important consequences for the emergence of new EHEC (Ahmed et al. 2008). Although virulence factors of the different pathotypes of *E. coli* normally do not overlap, there is an increasing number of studies describing *E. coli* isolates associated with human disease that contain new combinations of virulence factors.

(continued)

Inset 1.1 (continued)

EHEC is a subset of STEC, since the disease-causing ability of EHEC in humans is associated with its ability to express different types of Stx (the two immunologically distinct Stx1 and Stx2, and several variants of Stx2). The Stxs halt protein synthesis in the host cell, a process that may lead to an apoptotic cell death. Stx-mediated damage to renal glomerular endothelial cells is hypothesized as the precipitating event for HUS (Melton-Celsa et al. 2012). Nonetheless, EHEC is also a subset of EPEC, as it has the ability to form attaching and effacing (A/E) lesions on epithelial cells, a characteristic of EPEC. The capacity of EHEC to form A/E lesions in the intestine after ingestion is also linked to disease in humans. These A/E lesions are characterized by localized destruction (effacement) of brush border microvilli, attachment of the bacteria to the enterocyte membrane, and formation of a cup-like pedestal at the site of bacterial contact. The A/E lesion is mediated by the outer membrane adhesin intimin (encoded by the gene *eae*), its bacterially encoded receptor Tir, and effectors secreted through a type III secretion system. These proteins are encoded by the LEE (O'Sullivan et al. 2007; Melton-Celsa et al. 2012).

The 2011 European outbreak of EHEC O104:H4 infection was unusual in that there were important clinical and microbiological differences between this outbreak and previous large outbreaks, primarily those of STEC O157:H7 (Bielaszewska et al. 2011). The O104:H4 clone was not an entirely new clone but a slight variant of a known but seldom reported EHEC called HUSEC-41, which was first isolated in Germany in 2001 from a child with HUS (Mellmann et al. 2008). However, the *E. coli* strain causing the large outbreak in 2011 possesses an unusual combination of pathogenic features. It is resistant to β-lactam antibiotics due to an extended-spectrum β-lactamase, and it carries genes that are typically found in another group of diarrheagenic *E. coli*, the Enteroaggregative *E. coli* (EAEC) (Mellmann et al. 2011; Scheutz et al. 2011). In reality, EAEC is probably the most common bacterial cause of diarrhea but is not identified in most diagnostic laboratories. Comparative genomics suggests that the outbreak strain belongs to an EAEC lineage that acquired genes encoding the Stx2 toxin and antibiotic resistance (Mellmann et al. 2011; Scheutz et al. 2011).

This outbreak emphasizes the importance of being able to detect all pathogenic *E. coli* and not focusing on *E. coli* O157:H7 alone. Improved methods for identifying newly emergent strains of pathogenic *E. coli* are needed. The outbreak also highlights the importance of collaboration among all purveyors of public health to detect the outbreak, identify and characterize the causative agent, find the vehicles of transmission, and control the infection.

References

Ahmed N, Dobrindt U, Hacker J, Hasnain SE (2008) Genomic fluidity and pathogenic bacteria: applications in diagnostics, epidemiology and intervention. Nat Rev Microbiol 6:387–394.

Altekruse SF, Swerdlow DL (1996) The changing epidemiology of foodborne diseases. Am J Med Sci 311:23–29.

Ammon A, Makela P (2010) Integrated data collection on zoonoses in the EU, from animals to humans, and the analyses of the data. Int J Food Microbiol 30 Suppl 1:S43–S47.

Bielaszewska M, Mellmann A, Zhang W, Köck R, Fruth A, Bauwens A, Peters G, Karch H (2011) Characterisation of the *Escherichia coli* strain associated with an outbreak of haemolytic uraemic syndrome in Germany, 2011: a microbiological study. Lancet Infect Dis 11:671–676.

Blackburn CW, McClure PJ (2009) Foodborne pathogens: hazards, risk analysis and control, 2nd edn. Woodhead Publishing, Oxford, UK.

CDC (2011) Foodborne Diseases Active Surveillance Network (FoodNet): FoodNet surveillance report for 2010. http://www.cdc.gov/foodnet/reports.htm. Accessed 19 September 2011.

ECDC (2011) Shiga toxin-producing *Escherichia coli* (STEC): Update on outbreak in the EU (27 July 2011, 11:00). http://www.ecdc.europa.eu/en/activities/sciadvice/Lists/ECDC%20 Reviews. Accessed 12 September 2011.

EFSA (2011) The EU summary report on trends and sources of zoonoses, zoonotic agents and food-borne outbreaks in 2009. EFSA J 9:2090.

Feng P, Lampel KA, Karch H, Whittam TS (1998) Genotypic and phenotypic changes in the emergence of *Escherichia coli* O157:H7. J Infect Dis 177:1750–1753.

Feng P, Weagant SD (2011) Bacteriological Analytical Manual, Chapter 4A, Diarrheagenic *Escherichia coli*. http://www.fda.gov/food/scienceresearch/laboratorymethods/bacteriologicalanalyticalmanualbam/ucm070080.htm. Accessed 2 November 2011.

Ferens WA, Hovde CJ (2011) *Escherichia coli* O157:H7: animal reservoir and sources of human infection. Foodborne Pathog Dis 8:465–487.

Frank C, Werber D, Cramer JP, Askar M, Faber M, Heiden M, Bernard H, Fruth A, Prager R, Spode A, Wadl M, Zoufaly A, Jordan S, Kemper MJ, Follin P, Müller L, King LA, Rosner B, Buchholz U, Stark K, Krause G; HUS Investigation Team (2011) Epidemic profile of Shiga-toxin-producing *Escherichia coli* O104:H4 outbreak in Germany. N Engl J Med 365:1771–1780.

Green ST, Small MJ, Casman EA (2009) Determinants of national diarrheal disease burden. Environ Sci Technol 43:993–999.

Greig JD, Ravel A (2009) Analysis of foodborne outbreak data reported internationally for source attribution. Int J Food Microbiol 130:77–87.

Käferstein FK, Motarjemi Y, Bettcher DW (1997) Foodborne disease control: a transnational challenge. Emerg Infect Dis 3:503–510.

Karmali MA, Gannon V, Sargeant JM (2010) Verocytotoxin-producing *Escherichia coli* (VTEC). Vet Microbiol 140:360–370.

Koopmans M, Duizer E (2004) Foodborne viruses: an emerging problem. Int J Food Microbiol 90:23–41.

Kovats RS, Edwards SJ, Hajat S, Armstrong BG, Ebi KL, Menne B (2004) The effect of temperature on food poisoning: a time-series analysis of salmonellosis in ten European countries. Epidemiol Infect 132:443–453.

Kuchenmüller T, Hird S, Stein C, Kramarz P, Nanda A, Havelaar AH (2009) Estimating the global burden of foodborne diseases – a collaborative effort. Euro Surveill 14 pii:19195.

Lahuerta A, Westrell T, Takkinen J, Boelaert F, Rizzi V, Helwigh B, Borck B, Korsgaard H, Ammon A, Mäkelä P (2011) Zoonoses in the European Union: origin, distribution and dynamics – the EFSA-ECDC summary report 2009. Euro Surveill 16 pii:19832.

Lammerding AM, Fazil A (2000) Hazard identification and exposure assessment for microbial food safety risk assessment. Int J Food Microbiol 58:147–157.

Lefebvre B, Diarra MS, Moisan H, Malouin F (2008) Detection of virulence-associated genes in *Escherichia coli* O157 and non-O157 isolates from beef cattle, humans, and chickens. J Food Prot 71:1774–1784.

Mathusa EC, Chen Y, Enache E, Hontz L (2010) Non-O157 STEC in foods. J Food Prot 73:1721–1736.

Mellmann A, Bielaszewska M, Köck R, Friedrich AW, Fruth A, Middendorf B, Harmsen D, Schmidt MA, Karch H (2008) Analysis of collection of hemolytic uremic syndrome-associated enterohemorrhagic *Escherichia coli*. Emerg Infect Dis 14:1287–1290.

Mellmann A, Harmsen D, Cummings CA, Zentz EB, Leopold SR, Rico A, Prior K, Szczepanowski R, Ji Y, Zhang W, McLaughlin SF, Henkhaus JK, Leopold B, Bielaszewska M, Prager R, Brzoska PM, Moore RL, Guenther S, Rothberg JM, Karch H (2011) Prospective genomic characterization of the German enterohemorrhagic *Escherichia coli* O104:H4 outbreak by rapid next generation sequencing technology. PLoS One 6:e22751.

Melton-Celsa A, Mohawk K, Teel L, O'Brien A (2012) Pathogenesis of Shiga-toxin producing *Escherichia coli*. Curr Top Microbiol Immunol 357:67–103.

Newell DG, Koopmans M, Verhoef L, Duizer E, Aidara-Kane A, Sprong H, Opsteegh M, Langelaar M, Threfall J, Scheutz F, van der Giessen J, Kruse H (2010) Food-borne diseases – the challenges of 20 years ago still persist while new ones continue to emerge. Int J Food Microbiol 139 Suppl 1:S3–S15.

Nyachuba DG (2010) Foodborne illness: is it on the rise? Nutr Rev 68:257–269.

O'Sullivan J, Bolton D J, Duffy G, Baylis C, Tozzoli R, Wasteson Y, Lofdahl S (2007) Methods for detection and molecular characterisation of pathogenic *Escherichia coli*. Coordination action Food-CT-2006-036256. Pathogenic *Escherichia coli* network. http://www.antimicrobialresistance.dk/data/images/protocols/e%20coli%20methods.pdf. Accessed 3 November 2011.

Pennington H (2010) *Escherichia coli* O157. Lancet 376:1428–1435.

Pires SM, de Knegt L, Tine Hald T (2011) Estimation of the relative contribution of different food and animal sources to human *Salmonella* infections in the EU. Scientific/technical report submitted to EFSA. http://www.efsa.europa.eu/en/supporting/doc/184e.pdf. Accessed 17 October 2011.

Robert Koch Institute (2011) Informationen zum EHEC/HUS-Ausbruchsgeschehen – Ende des Ausbruchs [Information about the EHEC/HUS-outbreak– end of the outbreak]. http://www.rki.de/cln_117/nn_467482/DE/Content/InfAZ/E/EHEC/Info-HUS,templateId=raw,property=publicationFile.pdf/Info-HUS.pdf. Accessed 10 October 2011.

Roberts T (2007) WTP Estimates of the societal costs of U.S. food-borne illness. Am J Agric Econom 89:1183–1188.

Rotariu O, Ogden ID, Macritchie L, Forbes KJ, Williams AP, Cross P, Hunter CJ, Teunis PF, Strachan NJ (2011) Combining risk assessment and epidemiological risk factors to elucidate the sources of human *E. coli* O157 infection. Epidemiol Infect 27:1–16.

Scheutz F, Nielsen EM, Frimodt-Møller J, Boisen N, Morabito S, Tozzoli R, Nataro JP, Caprioli A (2011) Characteristics of the enteroaggregative Shiga toxin/verotoxin-producing *Escherichia coli* O104:H4 strain causing the outbreak of haemolytic uraemic syndrome in Germany, May to June 2011. Euro Surveill 16 pii:19889.

Sofos JN, Geornaras I (2010) Overview of current meat hygiene and safety risks and summary of recent studies on biofilms, and control of *Escherichia coli* O157:H7 in nonintact, and *Listeria monocytogenes* in RTE, meat products. Meat Sci 86:2–14.

Swaminathan B, Gerner-Smidt P (2007) The epidemiology of human listeriosis. Microbes Infect 9:1236–1243.

Tauxe RV (2002) Emerging foodborne pathogens. Int J Food Microbiol 78:31–41.

WHO (2011) Food safety. http://www.who.int/foodsafety/foodborne_disease/en/. Accessed 14 October 2011.

Chapter 2
Detection, Identification, and Analysis of Foodborne Pathogens

Abstract The detection and enumeration of microorganisms in food are an essential part of any quality control or food safety plan. Traditional methods of detecting food-borne pathogenic bacteria are often time-consuming because of the need for growth in culture media, followed by isolation, biochemical and/or serological identification, and in some cases, subspecific characterization. Advances in technology have made detection and identification faster, more sensitive, more specific, and more convenient than traditional assays. These new methods include for the most part antibody- and DNA-based tests, and modifications of conventional tests made to speed up analysis and reduce handling. With few exceptions, almost all assays used to detect specific pathogens in foods are qualitative assays, as they still lack sufficient sensitivity for direct testing and require some growth in an enrichment medium before analysis. One of the most challenging problems to circumvent with these assays is sample preparation. The possibilities of combining different rapid methods, including improved technologies for separation and concentration of specific bacteria, and for DNA extraction and purification, will facilitate the direct detection of pathogens in food. The goal is to avoid the enrichment, providing rapid alternatives to conventional quantitative culture methods. Further improvements, especially in genetic methods, can be expected, including the use of DNA microarray technology.

Keywords Detecting foodborne pathogens • Traditional culture methods • Rapid methods • Immunoassays • Molecular methods • Qualitative assays • Enrichment • Sample preparation

2.1 Introduction

The detection and enumeration of pathogens in food and on surfaces that come into contact with food are an important component of any integrated program to ensure the safety of foods throughout the food supply chain. Both government authorities and

G. López-Campos et al., *Microarray Detection and Characterization of Bacterial Foodborne Pathogens*, SpringerBriefs in Food, Health, and Nutrition, DOI 10.1007/978-1-4614-3250-0_2, © Guillermo Lopez-Campos, Joaquin V. Martinez-Suarez, Mónica Aguado-Urda, Victoria Lopez Alonso 2012

food companies use microbiological analysis to monitor the state of contamination at all times and analyze its trends so as to detect emerging risks. Microbiological analysis is also an essential tool for carrying out tests in accordance with the microbiological criteria established for each food type, as well as being essential for evaluating the actions of different management strategies based on the Hazard Analysis and Critical Control Points (HACCP) system (Stannard 1997; Jasson et al. 2010). The implementation of preventive systems such as the HACCP has greatly improved food safety, but it will not be fully effective until better methods of analysis are developed. These new detection methods are the necessary technologies that will substantially improve our food safety once integrated in the HACCP (Bhunia 2008).

Microbiological analysis of foods is based on the detection of microorganisms by visual, biochemical, immunological, or genetic means, either before enrichment (quantitative or enumerative methods) or after enrichment (qualitative methods, also known as presence/absence tests).

Traditional culture methods for detecting microorganisms in food are based on the incorporation of the food sample into a nutrient medium in which the microorganisms can multiply, thus providing visual confirmation of their growth. These conventional test methods are simple, easily adaptable, very practical, and generally inexpensive. Although not lacking in sensitivity, they can be laborious and depend on the growth of the microorganisms in different culture media (pre-enrichment, selective enrichment, selective plating, identification), which may require several days before results are known. Products that are minimally processed have an inherently short shelf life, which prevents the use of many of these conventional methods. Therefore, extensive research has been carried out over the years to reduce assay time through the use of alternative methods for detecting foodborne microorganisms and reduce the amount of manual labor by automating methods whenever possible (Jantzen et al. 2006a; Feng 2007; Betts and Blackburn 2009; Jasson et al. 2010).

In spite of its importance, the microbiological analysis of food has many limitations. Uncertainty of the analytical result must be considered when establishing microbiological criteria, including the variance associated with the sampling plan, method of analysis, and laboratory performance (Betts and Blackburn 2009). The microbiological analysis of food remains a challenging task for virtually all assays and technologies, especially for particular pathogenic species (Feng 2007). The problems may be due to

- The complexity of food matrices and composition.
- The heterogeneous distribution of low levels of pathogens.
- The stress suffered by the microorganisms during the processing of foods.
- The presence of bacteria from the normal microbiota, especially in raw foods.

The complexity of food matrices remains the major obstacle to the development of effective sampling and rapid testing methods (Feng 2007). Long duration enrichments are often used due to the low number of pathogenic microorganisms that tend to be present in food samples. Although previous enrichment is a limitation in terms of assay speed and precludes quantification of the original contaminant, it provides essential benefits, such as diluting the effects of inhibitors, allowing the differentiation of viable from nonviable cells, and allowing for the repair of cell stress or injury that may have resulted during food processing (Jantzen et al. 2006b; Wu 2008). Hence, it would be difficult to completely eliminate enrichment culture from the process of pathogen detection in foods (Feng 2007).

2.2   Separation and Concentration of Microorganisms Present in Food

The problem of low cell numbers present in food samples can also be solved by separating and concentrating microorganisms in food in order to discriminate the target pathogen from other cells and to use them at an appropriate concentration for the sensitivity level of the detection method. Food matrix materials can be simultaneously eliminated in order to avoid false-negative results. Several strategies including antibody-based as well as physical- and chemical-based methods have been developed for the separation and concentration of pathogens from various sample matrices (Stevens and Jaykus 2004; Bhunia 2008). In the case of beverages and liquid food, the concentration can be easily achieved through filtration or ultrafiltration (Chen et al. 2005; Hunter et al. 2011). However, these techniques do not allow the selective isolation of the organisms from a mixed population.

For solid foods, the main system present in the market is immunomagnetic separation and concentration (IMS) (Olsvik et al. 1994). With this technology, superparamagnetic particles or polystyrene beads are coated with iron oxide and antibodies that allow for the specific capture and isolation of intact pathogen cells present in suspensions of complex mixtures such as pre-enrichment media (Aminul Islam et al. 2006). The application of a magnetic field retains the particles along with the cells attached to the particles, allowing the rest of the organic and liquid material to be removed by washing. Captured antigens can be plated or further tested using other assays. IMS coupled with different rapid and automated assays has been used for the detection of pathogens such as *E. coli* O157:H7 (Seo et al. 1998; Fu et al. 2005; Aminul Islam et al. 2006; Hunter et al. 2011). In fact, IMS does not yield a pure culture of the target microbe, needing to be coupled to other tests for more definitive detection and identification results (Feng 2007).

There is an IMS technique that uses samples 500 times larger than most common assays by recirculating the sample during the capture phase in order to increase sensitivity and reduce detection times (Fedio et al. 2011). Specific bacteriophage tail–associated proteins can be attached to paramagnetic beads instead of antibodies as a way of capturing bacteria in suspension. There are bacteriophage-based capture kits that can be integrated into rapid detection methods in a similar way as in IMS (Favrin et al. 2003). The IMS may be employed either directly or after enrichment, but at present the major drawback of all these IMS-based assays is the susceptibility to inhibitors or interference by food components and the requirement for enrichment (Feng 2007).

2.3 Traditional Culture Methods

Standardized methods (e.g., ISO methods) are usually considered the reference analytical methods for official controls. In most cases, they are traditional culture methods that use selective liquid or solid culture media, to grow, isolate, and enumerate the target microorganism and simultaneously prevent the growth of other microorganisms present in the food (Jasson et al. 2010).

2.3.1 Quantitative Culture Methods

Enumeration of the microorganisms present in a sample is normally performed by plate count method or the most probable number (MPN) method. The plate count method is based on culturing dilutions of sample suspensions in the interior or on the surface of an agar layer in a Petri dish. Individual microorganisms or small groups of microorganisms will grow to form individual colonies that can be counted visually. The MPN method calculates the number of viable microorganisms in a sample by preparing decimal dilutions of the sample, and transferring subsamples of 3 serial dilutions to 9 or 15 tubes containing liquid culture medium, to carry out the method on 3 or 5 tubes, respectively. The tubes are incubated, and those that show growth (turbidity) are counted. Taking into account the dilution factor, the final result is compared to a standard MPN table, which will indicate the MPN of bacteria in the product (Blodgett 2010). This method is more labor-intensive and expensive than plate counting. The confidence limits are also quite large, even when studying many replica samples of each dilution level. The method is therefore usually less accurate than plate count methods but has the advantage of being more sensitive. Thus, it is widely used for estimations of levels of bacteria below 10 per gram of food (Table 2.1) (Stannard 1997; Betts and Blackburn 2009).

2.3.2 Qualitative Culture Methods

Qualitative procedures are used when it is not necessary to know the amount of a microorganism present in a sample but only its presence or absence. The technique requires an accurately weighed sample (usually 25 g). The typical colonies of the target microorganism on a selective/differential solid medium plate are often called *presumptive*. To confirm the identity of the desired microorganism, various biochemical and/or serological tests need to be carried out with pure cultures obtained from these presumptive colonies (Betts and Blackburn 2009) (see Sect. 2.6).

Table 2.1 Main characteristics of some culture-based and rapid detection methods (Adapted from Mandal et al. 2011)

Test method		Sensitivity[a]	Specificity	Duration of the assay (h)
Qualitative culture	Presence/absence	Defined by the quantity of food examined, e.g., presence in 25 g	Good	>72
Qualitative rapid detection	Presence/absence	Defined by the quantity of food examined, e.g., presence in 25 g	Variable[b]	Variable[b]
Quantitative culture	MPN	<10–100 MPN of bacteria per gram	Good	24–48
	Viable counts	>10–100	Good	24–72
	Impedance	100	Moderate/ good	6–24
Quantitative rapid detection	Bioluminescence	10^4	No[c]	<1–3
	DEFT, SPC	10^3–10^4	No[c]	<1
	Flow cytometry	10^5–10^7	Good	<1
	Immunological methods (LFD, ELISA, ELFA)	10^4–10^5	Moderate/ good	<1–3
	Nucleic acid–based assays (FISH, Q-PCR)	10^3–10^4	Excellent	<1–3

[a] CFU per g or mL, unless otherwise stated
[b] Depends on the rapid method used
[c] Can be made specific with specific separation and concentration steps (IMS- or phage-based) and/ or selective labeling of cells

2.4 Rapid and Automated Methods

The fast pace at which rapid methods are being developed precludes a discussion of all available methods. In this section, the breadth of rapid methods available and the scientific principles of the methods used for detection of pathogenic bacteria in foods are revised. Existing methods are presented in various formats and continue to be modified or adapted so that current methods have to be validated or evaluated using traditional standardized culture methods as reference (Feng 2007; Jasson et al. 2010; AOAC International 2011). Some rapid methods may also be considered as reference methods if they are shown to provide more accurate results than culture methods, such as current methods for detecting enterohemorrhagic strains of *E. coli* other than O157:H7, for example (Inset 2.1).

2.4.1 Changes to Culture Methods

Automation may be very useful in reducing the time required to prepare culture media, perform serial dilutions, count colonies, etc. (Fung et al. 1988; Jasson et al. 2010). There are a wide variety of rapid culture methods that have been designed to replace the standard agar plate, reducing the workload, facilitate rapid implementation, simplify handling, and/or reduce the need for a complete laboratory infrastructure, which do not necessarily shorten assay times. Some of these modified culture methods are based on the colony counting method, using, for instance, disposable cardboards containing dehydrated media (Chain and Fung 1991). Others are based on the MPN method (Torlak et al. 2008). In recent years a staggering number of chromogenic and fluorogenic culture media have been developed for the detection and enumeration of specific bacteria. The addition of these media to culturing protocols facilitates the rapid identification of presumptive colonies of the target microorganism (Manafi 2000). This has led to its incorporation in some official methods (Jasson et al. 2010).

The production of positively or negatively charged end products by bacteria during active growth (in initial stages of nutrient degradation) results in a variation in impedance of the culture medium. This can be measured at regular intervals over a period of 24 h after inoculation in specific media. This variation is proportional to the change in the number of bacteria in the culture. Thus, bacterial growth can be quantified (Wawerla et al. 1999; Jasson et al. 2010). This system is capable of analyzing hundreds of samples at a time since the instrument is computer-driven and automated to enable continuous monitoring. This technique is suited for testing samples with a low number of microorganisms. The limit of detection can be as low as 100 colony-forming units (CFU) per mL (Bosilevac et al. 2005) (Table 2.1). Methods based on the measure of impedance may be used in both quantitative and qualitative applications to detect all microorganisms or specific pathogens (Yang and Bashir 2008).

2.4.2   ATP Bioluminescence

This technique measures the emission of light produced by an enzymatic reaction between luciferin and luciferase that requires the presence of ATP (biolumines-cence). The amount of light produced (measured by means of a luminometer) is proportional to the concentration of ATP, and therefore the number of microorganisms in the original sample. Bioluminescence produced by ATP may be used to enumerate the total microorganisms in a sample but is only applicable if the number of bacteria present is high (more than 10^4 CFU/g) (Samkutty et al. 2001; Jasson et al. 2010) (Table 2.1). The technique is widely used to measure the cleanliness of surfaces that come into contact with food, including the presence of organic residues and microbial contaminants. It provides results in less than 5 min (Cunningham et al. 2011). The system lacks specificity, however. An alternative and more specific approach includes an IMS step for capturing the target bacteria, which is then detected by bioluminescence (Hunter et al. 2011); species-specific bacteriophages instead of chemicals can be used to lyse cells to release ATP, thereby providing additional assay specificity (Kannan et al. 2010).

2.4.3   Microscopic Methods

2.4.3.1   Direct Epifluorescent Filter Technique

The direct epifluorescent filter technique (DEFT) is a microscopic method for the enumeration of viable cells in a sample based on the binding properties of the fluorochrome acridine orange. Once treated with detergents and proteolytic enzymes, the samples are filtered through a polycarbonate membrane. The cells are stained on this same filter and examined under an epifluorescent microscope (Pettipher et al. 1992), a process that can be carried out semi-automatically by connecting the microscope to an image analysis system (Hermida et al. 2000). The number of viable cells can be obtained in 10 min. However, DEFT is a very labor-intensive technique that does not have the capability of processing a large number of samples and is only applicable if the number of bacteria present is high (10^3–10^4 CFU/g) (Table 2.1). Additionally, fluorescent food material can be trapped on the filter, and the technique can only be used with raw food and usually for enumerating total viable microorganisms (Hermida et al. 2000). Nevertheless, DEFT may be used for the detection and enumeration of specific bacteria in food samples provided they can be isolated from the unfilterable matrix. IMS followed by DEFT and solid-phase cytometry (see Sect. 2.4.3.3) have given results that compared favorably with IMS followed by plating (see below).

2.4.3.2   Flow Cytometry

Flow cytometry quantitatively measures optical characteristics of cells when they are forced to pass individually through a beam of light. Fluorescent dyes can be used to test the viability and metabolic state of microorganisms (Veal et al. 2000). Samples are injected into a fluid (dye), which passes through a sensing medium in a flow cell. The cells are carried by the laminar flow of water through a focus of light, each cell emits a pulse of fluorescence, and the scattered light is collected by lenses and directed onto selective detectors (photomultiplier tubes). This technique is fast, automatic, and potentially very specific, as long as appropriate dyes are available for selectively labeling specific types of microorganisms and appropriate methods for separating cells from food are utilized so as not to interfere with detection (Seo et al. 1998). The sensitivity of flow cytometry, however, is low (Table 2.1); the detection limit with food samples is around 10^5–10^7 CFU/g (Betts and Blackburn 2009). Currently, there are various flow cytometry methods developed for foods, especially for liquid samples such as dairy products, water, and other beverages (Comas-Riu and Rius 2009).

2.4.3.3   Solid-Phase Cytometry

Solid-phase cytometry (SPC) is a technique that combines aspects of flow cytometry and epifluorescence microscopy (D'Haese and Nelis 2002). After filtration of the sample, the retained microorganisms are fluorescently labeled with argon laser excitable dyes on the membrane filter and automatically counted by a laser scanning device. Each fluorescent spot can be visually inspected with an epifluorescence microscope connected to a scanning device by a computer-driven moving stage. Depending on the fluorogenic labels used, information on the identity and the physiological status of the microorganisms can be obtained within a few hours. SPC, like DEFT (see Sect. 2.4.3.1), is only applicable if the number of bacteria present is high (10^3–10^4 CFU/g). Although both techniques were originally recommended for the determination of the total viable microbial count in liquid samples, they may also be utilized for the rapid detection and enumeration of pathogens in food samples, provided they can be specifically isolated. The efficiency of viable bacteria detection from foods by IMS followed by SPC and DEFT was assessed using the pathogen *E. coli* O157:H7 (Pyle et al. 1999). Within 5–7 h of enrichment, the IMS-SPC method detected higher numbers of cells than were detected by plating. SPC in conjunction with fluorescent viability staining has also been reported as a tool to detect viable but nonculturable *Campylobacter jejuni* (Cools et al. 2005).

2.4.4   Immunological Detection Methods

The antibody-based system has facilitated the design of a variety of assays and formats. In some cases, the antigen–antibody complex formed is directly measurable or even visible. Incubation times are usually very short for methods such as agglutination reactions commonly used for the rapid identification of microorganisms (see Sect. 2.6). Normally, the antibody is labeled with a fluorescent reagent or with an enzyme so that the antigen–antibody interaction may be visualized more easily when it occurs.

2.4.4.1   Lateral Flow Devices

Lateral flow devices (LFD) are typically comprised of a simple dipstick made of a porous membrane that contains colored latex beads or colloidal gold particles coated with detection antibodies targeted toward a specific microorganism. The particles are found on the base of the dipstick, which is put in contact with the enrichment medium (Posthuma-Trumpie et al. 2009). If the target organism is present, then it will bind with the colored particles. This conjugated cell/particle moves by capillary action until it finds the immobilized capture antibodies. Upon binding with these, it forms a colored line that is clearly visible in the device window, indicating a positive result (Betts and Blackburn 2009). As with other immunoassays, LFD also require previous enrichment. The technique is extremely simple to use and easy to interpret, requires no washing or manipulation, and can be completed within 10 min after culture enrichment (Aldus et al. 2003). There are various LFD on the market that have been validated for detecting different foodborne pathogens (Jasson et al. 2010; AOAC International 2011).

2.4.4.2   Enzyme-Linked Immunosorbent Assay and Enzyme-Linked Fluorescence Assay

The enzyme-linked immunosorbent assay (ELISA) is a biochemical technique that combines an immunoassay with an enzymatic assay. As LFD, it is a "sandwich" assay. An antibody bound to a solid matrix is used to capture the antigen from enrichment cultures and a second antibody conjugated to an enzyme is used for detection. The enzyme is capable of generating a product detectable by a change in color, or in the case of enzyme-linked fluorescence assay (ELFA) in fluorescence, which allows for indirect measurement using spectrophotometry (or fluorometry for ELFA) of the antigen present in the sample (microorganism or toxin) (Cohen and Kerdahi 1996; Jasson et al. 2010).

Detection using automated and robotic ELISAs is widely used since they can reduce detection times after enrichment to as low as 1–3 h (Thacker et al. 1996). Thus, the results can be obtained in 2–3 days instead of the 3–5 days needed by conventional methods (Leon-Velarde et al. 2009). There are many commercial enzyme immunoassays for detecting the main pathogens and toxins in foods (AOAC International 2011). Bacteriophage recombinant protein technology can also be integrated in detection methods as part of improved immunological qualitative tests (Jasson et al. 2010; Savoye et al. 2011) (Inset 2.1).

The success of an immunoassay depends on the specificity of the antibody. Using hybridoma technology, it has been possible to develop monoclonal antibodies that react only with one specific pathogen. The limit of detection for immunoassays is approximately 10^4–10^5 CFU/g (Table 2.1) depending on the type of antibody and its affinity for the corresponding epitope, which means that one or two previous enrichment stages are always required (Jasson et al. 2010).

2.4.5   Molecular Detection Methods

There has been an explosion in the past 15 years in the introduction of nucleic acid–based assays for the detection and identification of foodborne pathogens. There are many DNA-based assay formats, but only probes and nucleic acid amplification techniques have been developed commercially for detecting foodborne pathogens.

2.4.5.1   Fluorescent In Situ Hybridization

Fluorescent in situ hybridization (FISH) with oligonucleotide probes directed at rRNA is the most common method among molecular techniques not based on PCR. The probes used by FISH tend to be 15–25 nucleotides in length, and are covalently labeled at their 5' end with fluorescent labels. After hybridization, the specifically stained cells are detected using epifluorescence microscopy (Wagner et al. 2003). The detection limit of this technique is around 10^4 CFU/g (Table 2.1). Following pre-enrichment to reach these detection levels, the results can be obtained quickly (in about 3 h) (Bottari et al. 2006). FISH in combination with flow cytometry has been used for rapid culture-independent detection of *Salmonella* spp. on the surfaces of tomatoes and other fresh produce (Bisha and Brehm-Stecher 2010).

2.4.5.2 Polymerase Chain Reaction

Polymerase chain reaction (PCR) is a method used for the in vitro enzymatic synthesis of specific DNA sequences by *Taq* and other thermoresistant DNA polymerases. PCR uses oligonucleotide primers that are usually 20–30 nucleotides in length and whose sequence is homologous to the ends of the genomic DNA region to be amplified. The method is performed in repeated cycles, so that the products of one cycle serve as the DNA template for the next cycle, doubling the number of target DNA copies in each cycle (Hill 1996). The rapid increase in the number of copies of the target sequence that can be achieved with PCR-based methods makes them ideal candidates for the development of faster microbiological detection systems. Many PCR tests have been validated and commercialized to make PCR a standard tool used by food microbiology laboratories to detect pathogens in foods (Jasson et al. 2010; AOAC International 2011).

Conventional PCR relies on amplification of the target gene(s) in a thermocycler, separation of PCR products by gel electrophoresis, followed by visualization and analysis of the resulting electrophoretic patterns, a process that can take a number of hours. The specificity can be subsequently confirmed by sequencing the amplified fragment. PCR can be superior to culture for detecting the main pathogens in food samples (Abubakar et al. 2007).

Real-time PCR allows both the detection and quantification of a signal emitted by the amplified product by using the continuous measurement of a fluorescent label during the PCR reaction. The increase in fluorescence can be monitored in real time, which allows accurate quantification over several orders of magnitude of the DNA target sequence. Results can be obtained in an hour or less, which is considerably faster than conventional PCR. Real-time PCR has greatly increased the speed and sensitivity of PCR-based detection methods (Hanna et al. 2005). But the terms "rapid" and "sensitive," when applied to PCR as a detection method for foodborne pathogens, must be used with caution (Jasson et al. 2010). PCR itself requires only about 30–90 min, but all methods for detecting foodborne pathogens using PCR require pre-enrichment times that may vary from 6–8 to 48 h. Aside from that, if a positive result for PCR is reached, this must be confirmed using cultures. Regarding sensitivity, the limit of quantification of real-time PCR with food samples is around 10^3–10^4 CFU/g (Table 2.1) (Rodriguez-Lazaro and Hernandez 2006; Navas et al. 2006; Jasson et al. 2010). This limit is still too high for quantitative detection, since most samples taken from throughout the food supply chain are usually contaminated with fewer pathogen cells (normally less than 100 CFU/g). Consequently, this requires some enrichment of the microorganisms to be done prior to analysis, which turns all currently available commercial real-time PCR into qualitative detection methods (presence/absence) instead of quantitative.

In some cases, multiple different microorganisms may be detected in a single PCR reaction by amplifying the corresponding loci simultaneously. In this type of multiplex PCR reaction, all necessary primers are combined in a single tube for detecting the presence of the main pathogens associated with a given food (Kawasaki et al. 2009) or the main subtypes within a given species (Valadez et al. 2011).

2.5 Conclusions About Traditional and Rapid Methods

The process of selecting an appropriate method must consider the main criteria of the sensitivity of analysis, the time of detection, and the specificity of the test (Table 2.1). The cornerstone of any method is its accuracy. This consists of the sensitivity and the specificity. The intent in developing a rapid assay is to reduce the time required to obtain an accurate result.

Qualitative detection (presence/absence) tests are used if information concerning the presence of an organism in a specified quantity of food is required. The sensitivity of these tests is then defined by the quantity of food examined (Stannard 1997; Jasson et al. 2010). In many cases, the requirement of detection is less than one cell per 25 g of food, as small numbers of some pathogens may be sufficient to cause disease. Sensitive quantitative detection is usually achieved by traditional culture methods (Table 2.1). MPN determinations are suitable for low counts such as less than 100 per gram, and are widely used for estimates of levels below 10 per gram. Plate counts are generally used for counts of more than 10 CFU/g but are more accurate when levels exceed 100 CFU/g. With rapid methods, the lower limit of detection is almost always above 10^3 CFU/g of food (Table 2.1). Thus, rapid methods still lack sufficient sensitivity for direct testing (Feng 2007).

Traditional culture methods may require many days, resulting in very long assay durations. Most rapid methods for the detection of pathogens or toxins can be done in a few minutes to a few hours or at the utmost 1 day (Table 2.1). However, many detection systems need an enrichment, and positive results must be confirmed by the appropriate official method, which involve culturing, in many instances (Feng 2007). In spite of this, commercially available rapid detection methods, such as ELISA, LFD, and PCR, have substantially shortened the total time of the detection assay when compared to conventional methods (Table 2.1) (Leon-Velarde et al. 2009). They are, therefore, of great use in the rapid analysis of food with the goal of ensuring that only negative samples or lots are sent to market (Bohaychuk et al. 2005). A major disadvantage of alternative methods over culture methods is that most rapid methods involve damaging the cells. Therefore, viable cells for confirmation and further characterization can only be obtained by carrying out repeat analyses using standard culture procedures (Feng 2007).

Traditional culture methods use selective liquid or solid culture media to grow the target microorganism and simultaneously prevent the growth of other microorganisms present in the food. The selection of a specific DNA sequence that will serve as probe or primer, along with the conditions in which hybridization is carried out, will determine the specificity of the nucleic acid–based assays. Concerns with immunological techniques include problems with cross-reactivity and difficulties with obtaining species-specific assays selecting appropriate antibodies.

Although the method selected may be rapid and accurate, other factors, including the speed of sample processing and cost, must be considered. Sample preparation limits the speed of the assay and is one of the most challenging problems for the direct detection of bacteria in food. To be able to calculate the actual number of bacteria present in a food by rapid methods such as quantitative PCR (Q-PCR), no previous enrichment of the sample can be performed. If the target is present in low numbers and a small volume of sample is taken (as many PCR methods only require 0.1 mL or less), there is a chance that this subsample may not include the target organism. Thus, a labor-intensive preparation of the sample is needed in order to recover food cells and quantitatively extract the DNA and purify it. The major bottleneck of Q-PCR is therefore found in the preparation of the sample (Jasson et al. 2010). The potential application of biosensor technology to pathogen testing in foods offers many attractive features (Chap. 6). However, like most assays, the exquisite sensitivity achievable with cultures does not translate to food testing (Feng 2007). Microbial sensors are particularly applicable in fluid systems with little organic substances, but this technique can present problems in its efficacy in food systems containing fats and proteins that coat the sensor and render it inoperable. Therefore, adequate sample preparation techniques are another important consideration in developing biosensor assays for foods. The difficulties inherent in food testing apply to microarrays as well, since the sensitivity of these assays decreases when testing foods (Chap. 5). Hence, factors such as adequate sample preparation and the need for confirmation of results must be considered in designing microarrays for food testing (Feng 2007). More research is needed on techniques for separating microorganisms from the food matrix and for concentrating them before detection methods are carried out. The possibilities of combining different rapid methods, including improved technologies for separation and concentration of specific bacteria, and for DNA extraction and purification (Chap. 5), will facilitate the direct detection of pathogens in food (Mandal et al. 2011). The goal is to avoid the enrichment providing rapid alternatives to conventional quantitative culture methods.

2.6 Identification and Characterization of Microorganisms

After detection, miniature biochemical kits are often used to identify microorganisms quickly and easily. Advances in instrumentation have enabled automation of identification tests. These instruments can incubate the reactions and automatically monitor biochemical changes to generate a phenotypic profile, which is then compared with a database stored in the computer to provide an identification (Stager and Davis 1992). Other automated systems identify bacteria based on compositional or metabolic properties, such as fatty acid profiles, carbon oxidation profiles, or other traits (Miller and Rhoden 1991).

It is also very common to use antibodies to detect specific antigens using simple agglutination reactions. There are several commercial methods based on agglutination, in which antibody-coated colored latex beads or colloidal gold particles are used for quick confirmation or serological identification of pure culture isolates of bacteria from foods (D'Aoust et al. 1991; van Griethuysen et al. 2001). A modification of latex agglutination, known as reverse passive latex agglutination (RPLA), tests for soluble antigens and is used mostly in testing for toxins in food extracts or for toxin production by pure cultures (Feng 2007).

Molecular methods based on the hybridization or amplification of nucleic acids may also be used to identify or confirm the identities of microorganisms, as well as a subspecific characterization. Molecular typing of a species can help not only to investigate the origin of the strains present in foods but also to establish an association of the various degrees of virulence (López et al. 2006) or antimicrobial resistance that may exist within a species to certain strains or subtypes. There are plenty of molecular genotyping and subtyping methods. Currently, the most widely used technique for microbial source tracking is pulsed-field gel electrophoresis (PFGE), which is based in macrorestriction analysis of bacterial DNA and usually is carried out in reference or public health laboratories (Foley et al. 2009). Ribotyping is the only molecular typing technique that has been marketed in a completely automated format and allows for the large-scale characterization and fingerprinting of strains for epidemiological investigation. Ribotyping is a variant of the restriction fragment length polymorphism (RFLP) technique, which employs probes based on rDNA. Results can be obtained within 16 h, and riboprint patterns can be stored to create a unique database (Pavlic and Griffiths 2009).

Inset 2.1 Detection of Enterohemorrhagic Strains of *Escherichia coli* in Food

Enterohemorrhagic *E. coli* (EHEC) is a subgroup of Shiga-toxin- (Stx) producing *E. coli* (STEC) that has caused major outbreaks and sporadic cases of hemorrhagic colitis (HC) and hemolytic uremic syndrome (HUS) (Inset 1.1). Among EHEC, *E. coli* O157:H7 is now recognized as an important human pathogen and together with *Salmonella, L. monocytogenes,* and *Campylobacter* is one of four major food safety parameters (Jasson et al. 2010).

There are other well-known EHEC serotypes that have caused illness worldwide, representing a growing public health concern. Some cases of non-O157 STEC illness appear to be as severe as cases associated with O157. One of the more recent examples of severe disease caused by non-O157 EHEC is the

(continued)

Inset 2.1 (continued)

2011 European outbreak of EHEC O104:H4 infection (Inset 1.1). Nevertheless, most cases of STEC attributed to non-O157 are less severe. There is much variation in virulence potential within STEC serotypes, and many may not be pathogenic. Of the more than 400 serotypes isolated, fewer than 10 (O26, O45, O91, O103, O111, O113, O121, O128, and O145) cause the majority of all STEC-related human illnesses (Mathusa et al. 2010).

Detection of *E. coli* O157:H7

Because of the low dose of *E. coli* O157:H7 needed to cause infection, sensitive and rapid detection methods for *E. coli* O157:H7 in food samples are necessary in order for the food industry to ensure a safe supply of foods. The sensitive detection of *E. coli* O157:H7 has been developed over recent years with the use of selective enrichment and through the development of methods such as IMS. Selective enrichment followed by IMS and subsequent spread plating of the concentrated target cells onto selective and differential agar appears to be the most sensitive and cost-effective method for the isolation of *E. coli* O157 from raw foods (Bolton et al. 1996).

There are different reference methods for detecting *E. coli* O157:H7 in foods such as the protocol ISO 16654 (International Organization for Standardization 2001), the USDA-FSIS method used for the analysis of both raw and ready-to-eat meat products and environmental samples (United States Department of Agriculture Food Safety and Inspection Service 2010a, b), and the FDA-BAM screening method for other foods (Feng and Weagant 2011).

Currently, there are at least 22 commercial assays available for the detection and/or identification of *E. coli* O157:H7 that have been officially validated for use in food testing (AOAC International 2011). These rapid methods are mainly based on the use of antibodies or PCR.

Detection of Non-O157 EHEC

Except for O157:H7, there are no clear regulations to address the presence of other EHEC strains in foods. This is partly due to the difficulties in discerning EHEC from STEC strains that have not been implicated in illness and may not be pathogenic (Mathusa et al. 2010). These strains have few reliable biochemical or morphological characteristics (besides Stx production) that allow them to be distinguished from commensal *E. coli*. Thus, to detect EHEC other than O157 and phenotypic variants of *E. coli* O157 in food, methods for the detection of virulence factors and/or genes must be used.

While there is no standardized protocol for non-O157 EHEC, IMS has been shown to be useful in the recovery of specific serogroups from food and

(continued)

Inset 2.1 (continued)

---

fecal samples (O'Sullivan et al. 2007). Together with O157, paramagnetic beads coated with antibodies against serogroups O26, O103, O111, and O145 are also commercially available. There are also commercial selective agars that differentiate among O157 and those four EHEC serogroups on the basis of color, which is determined by a chromogenic compound and a mixture of selected carbohydrates (O'Sullivan et al. 2007).

Real-time PCR assays have been developed for the detection of EHEC that carry the major associated virulence genes *eae*, *stx1*, and *stx2* (Inset 1.1) and/or serogroup-specific gene targets. Serogroup target genes include *wzx* (O-antigen flippase in O26), *wzy* (O-antigen polymerase in O103), *galE* (galactose operon in O103), *Wbdl* (transferase gene in O111), and *Sil* (silver resistance in O145) (O'Sullivan et al. 2007). Some of these targets have been combined into multiplex assays (O'Sullivan et al. 2007; Valadez et al. 2011). Unfortunately, when confirmation of the isolates inoculated in food samples is carried out, the recovery of some serogroups was lower than that of others (Fratamico et al. 2011), and for serogroup O111, false negatives were often found (Verstraete et al. 2012). In the case of the 2011 European outbreak of EHEC O104:H4 infection, simple diagnostic screening tools to detect the outbreak strain in clinical specimens and foods were rapidly obtained. Three days after having the first isolate, a specific multiplex PCR was developed and made public, allowing for the specific identification of the strain, together with a novel real-time PCR assay for its detection in foods (Scheutz et al. 2011).

The USDA-FSIS procedure to detect and isolate non-O157 STEC serogroups utilizes a multiplex real-time PCR detection assay followed by culture isolation (United States Department of Agriculture Food Safety and Inspection Service 2010c). The culture isolation of non-O157 EHEC involves IMS followed by plating onto commercial chromogenic agars. Typical colonies from these plates are confirmed using multiplex PCR assays and biochemical identification. In the FDA method, the enrichment procedure and real-time PCR screening assay for O157 STEC have also been validated for the detection and recovery of other non-O157 EHEC as well (Feng and Weagant 2011). Commercial real-time PCR assays have also been designed for each of the non-O157 STEC O serogroups most commonly associated with human illness (Lin et al. 2011).

Identification and Characterization of EHEC

Serotyping is based on the use of specific antisera and the detection of O- and H-antigens expressed by these bacteria. Currently, a total number of 181 O-antigens and 53 H-antigens are available. Kits for O serotyping and H serotyping for STEC are commercially available. However, a minority of strains

(continued)

Inset 2.1 (continued)

do not serotype satisfactorily (O'Sullivan et al. 2007). Serotyping can also be done by molecular methods including analysis of gene/DNA sequences within and just outside the O-antigen cluster and *fliC* (H-antigens) sequence analysis. This molecular serotyping is based on DNA sequencing, DNA hybridization (microarray), PCR, or PCR-RFLP (O'Sullivan et al. 2007). The importance of serotyping was highlighted when during the 2011 European outbreak of EHEC O104:H4, other HUS-associated EHEC serogoups (O157, O91, and O103) were also identified, giving misleading results. This shows the need for the development of rapid serotyping and pathotyping methods for all HUS-associated *E. coli* strains (Friedrich 2011).

Genes encoding variants of Stx, intimin, enterohemolysins, and other potential pathogenicity factors (Inset 1.1) can be used as targets for characterization and typing. Single-nucleotide polymorphism (SNP) assays detect nucleotide substitutions (SNPs) by using sequencing, PCR, or microarrays. While *stx1* has a relatively conserved nucleotide sequence, several variants of *stx2* have been described (O'Sullivan et al. 2007). Neupane et al. (2011) have shown that overexpression of *stx2* is common in strains associated with HUS and that SNPs that may affect *stx2* expression could be useful in differentiating between highly virulent strains.

There are numerous PCR-based methods for the detection of EHEC virulence factors, but the time and cost involved with large-scale screening efforts and population level analyses have limited the size and scope of studies. Combining the high-throughput performance of microarrays with the specificity and sensitivity of real-time Q-PCR, Gonzales et al. (2011) identified and evaluated a panel of 28 genetic markers that can be used with high-throughput PCR to virulotype, serotype, and preliminarily subtype large numbers of isolates. The panel includes known virulence and regulatory genes, O-antigen genes, and select prophage regions of O157 and non-O157 EHEC. This has the potential to become an integral tool in outbreak, environmental, and genetic investigations of EHEC.

A different microarray has been developed for the identification of STEC strains with a high potential for human virulence. It is based on the genetic identification of 12 O-types and 7 H-types of STEC including the most clinically relevant EHEC serotypes (Bugarel et al. 2010). The genes selected for determination of the O-antigens showed a high specificity and concordance with serology. The microarray also had a high specificity for EHEC-associated virulence factors, and it was used during the 2011 European outbreak of EHEC O104:H4 infection (Chap. 5).

References

Abubakar I, Irvine L, Aldus CF, Wyatt GM, Fordham R, Schelenz S, Shepstone L, Howe A, Peck M, Hunter PR (2007) A systematic review of the clinical, public health and cost-effectiveness of rapid diagnostic tests for the detection and identification of bacterial intestinal pathogens in faeces and food. Health Technol Assess 11:1–216.

Aldus CF, van Amerongen A, Ariens RMC, Peck MW, Wichers JH, Wyatt GM (2003) Principles of some novel rapid dipstick methods for detection and characterization of verotoxigenic *Escherichia coli*. J Appl Microbiol 95:380–389.

Aminul Islam M, Heuvelink AE, Talukder KA, de Boer E (2006) Immunoconcentration of Shiga toxin-producing *Escherichia coli* O157 from animal faeces and raw meats by using Dynabeads anti-*E. coli* O157 and the VIDAS system. Int J Food Microbiol 109:151–156.

AOAC INTERNATIONAL (2011) Performance Tested Methods^sm Validated Methods. http://www.aoac.org/testkits/testedmethods.html. Last updated: October 2011. Accessed 18 November 2011.

Betts R, Blackburn CW (2009) Detecting pathogens in food. In: Foodborne pathogens: hazards, risk analysis and control, 2nd edn. Edited by: Blackburn CW, McClure PJ. Woodhead Publishing, Oxford, UK. pp. 17–65.

Bhunia AK (2008) Biosensors and bio-based methods for the separation and detection of food-borne pathogens. Adv Food Nutr Res 54:1–44.

Bisha B, Brehm-Stecher BF (2010) Combination of adhesive-tape-based sampling and fluores-cence *in situ* hybridization for rapid detection of *Salmonella* on fresh produce. J Vis Exp 44 pii:2308.

Blodgett R (2010) Most probable number from serial dilutions. Bacteriological Analytical Manual Appendix2. http://www.fda.gov/Food/ScienceResearch/LaboratoryMethods/Bacteriological AnalyticalManualBAM/ucm109656.htm. Accessed 10 November 2011.

Bohaychuk VM, Gensler GE, King RK, Wu JT, McMullen LM (2005) Evaluation of detection methods for screening meat and poultry products for the presence of foodborne pathogens. J Food Prot 68:2637–2647.

Bolton FJ, Crozier L, Williamson JK (1996) Isolation of *Escherichia coli* O157 from raw meat products. Lett Appl Microbiol 23:317–321.

Bosilevac JM, Shackelford SD, Brichta DM, Koohmaraie M (2005) Efficacy of ozonated and electrolyzed oxidative waters to decontaminate hides of cattle before slaughter. J Food Prot 68:1393–1398.

Bottari B, Ercolini D, Gatti M, Neviani E (2006) Application of FISH technology for microbio-logical analysis: current state and prospects. Appl Microbiol Biotechnol 73:485–494.

Bugarel M, Beutin L, Martin A, Gill A, Fach P (2010) Micro-array for the identification of Shiga toxin-producing *Escherichia coli* (STEC) seropathotypes associated with Hemorrhagic Colitis and Hemolytic Uremic Syndrome in humans. Int J Food Microbiol 142:318–329.

Chain VS, Fung DYC (1991) Comparison of Redigel, Petrifilm, spiral plate system, Isogrid, and aerobic plate count for determining the numbers of aerobic bacteria in selected foods. J Food Prot 54:208–211.

Chen WT, Hendrickson RL, Huang CP, Sherman D, Geng T, Bhunia AK, Ladisch MR (2005) Mechanistic study of membrane concentration and recovery of *Listeria monocytogenes*. Biotechnol Bioeng 89:263–273.

Cohen AE, Kerdahi KF (1996) Evaluation of a rapid and automated enzyme-linked fluorescent immunoassay say for detecting *Escherichia coli* serogroup O157 in cheese. J AOAC Int 79:858–860.

Comas-Riu J, Rius N (2009) Flow cytometry applications in the food industry. J Ind Microbiol Biotechnol 36:999–1011.

Cools I, D'Haese E, Uyttendaele M, Storms E, Nelis HJ, Debevere J (2005) Solid phase cytometry as a tool to detect viable but non-culturable cells of *Campylobacter jejuni*. J Microbiol Methods 63:107–114.

Cunningham AE, Rajagopal R, Lauer J, Allwood P (2011) Assessment of hygienic quality of surfaces in retail food service establishments based on microbial counts and real-time detection of ATP. J Food Prot 74:686–690.

D'Aoust J-Y, Sewell AM, Greco P (1991) Commercial latex agglutination kits for the detection of foodborne *Salmonella*. J Food Prot 54:725–730.

D'Haese E, Nelis HJ (2002) Rapid detection of single cell bacteria as a novel approach in food microbiology. J AOAC Int 85:979–983.

Favrin SJ, Jassim SA, Griffiths MW (2003) Application of a novel immunomagnetic separation-bacteriophage assay for the detection of *Salmonella* Enteritidis and *Escherichia coli* O157: H7 in food. Int J Food Microbiol 85:63–71.

Fedio WM, Jinneman KC, Yoshitomi KJ, Zapata R, Wendakoon CN, Browning P, Weagant SD (2011) Detection of *E. coli* O157:H7 in raw ground beef by Pathatrix™ immunomagnetic-separation, real-time PCR and cultural methods. Int J Food Microbiol 148:87–92.

Feng P (2007) Rapid methods for the detection of foodborne pathogens: current and next-generation technologies. In: Food microbiology, fundamentals and frontiers, 3rd edn. Edited by: Doyle MP, Beuchat LR. ASM Press, Washington, D.C. pp 911–934.

Feng P, Weagant SD (2011) Diarrheagenic *Escherichia coli*. Bacteriological Analytical Manual, Chapter 4A. http://www.fda.gov/food/scienceresearch/laboratorymethods/bacteriologicalanalyticalmanualbam/ucm070080.htm. Accessed 10 November 2011.

Foley SL, Lynne AM, Nayak R (2009) Molecular typing methodologies for microbial source tracking and epidemiological investigations of Gram-negative bacterial foodborne pathogens. Infect Genet Evol 9:430–440.

Fratamico PM, Bagi LK, Cray WC Jr, Narang N, Yan X, Medina M, Liu Y (2011) Detection by multiplex real-time polymerase chain reaction assays and isolation of Shiga toxin-producing *Escherichia coli* serogroups O26, O45, O103, O111, O121, and O145 in ground beef. Foodborne Pathog Dis 8:601–607.

Friedrich AW (2011) Enterohaemorrhagic *Escherichia coli* O104:H4: are we prepared now? Euro Surveill 16 pii:19938.

Fu Z, Rogelj S, Kieft TL (2005) Rapid detection of *Escherichia coli* O157: H7 by immunomagnetic separation and real-time PCR. Int J Food Microbiol 99:47–57.

Fung DYC, Cox NA, Bailey JS (1988) Rapid methods and automation in the microbiological examination of food. Dairy Food Sanit 8:292–296.

Gonzales TK, Kulow M, Park DJ, Kaspar CW, Anklam KS, Pertzborn KM, Kerrish KD, Ivanek R, Döpfer D (2011) A high-throughput open-array qPCR gene panel to identify, virulotype, and subtype O157 and non-O157 enterohemorrhagic *Escherichia coli*. Mol Cell Probes 25:222–230.

Hanna SE, Connor CJ, Wang HH (2005) Real-time polymerase chain reaction for the food microbiologist: technologies, applications, and limitations. J Food Sci 70:R49–R53.

Hermida M, Taboada M, Menéndez S, Rodríguez-Otero JL (2000) Semi-automated direct epifluorescent filter technique for total bacterial count in raw milk. J AOAC Int 83:1345–1348.

Hill WE (1996) The polymerase chain reaction: application for the detection of foodborne pathogens. CRC Crit Rev Food Sci Nutrit 36:123–173.

Hunter DM, Leskinen SD, Magaña S, Schlemmer SM, Lim DV (2011) Dead-end ultrafiltration concentration and IMS/ATP-bioluminescence detection of *Escherichia coli* O157:H7 in recreational water and produce wash. J Microbiol Methods 87:338–342.

International Standards Organization (2001) Microbiology of food and animal feeding stuffs – horizontal method for the detection of *Escherichia coli* O157. ISO 16654.

Jantzen MM, Navas J, Corujo A, Moreno R, Lopez V, Martinez-Suarez JV (2006a) Review. Specific detection of *Listeria monocytogenes* in foods using commercial methods: from chromogenic media to real-time PCR. Span J Agric Res 4:235–247.

Jantzen MM, Navas J, de Paz M, Rodríguez B, da Silva WP, Nuñez M, Martínez-Suárez JV (2006b) Evaluation of ALOA plating medium for its suitability to recover high pressure-injured *Listeria monocytogenes* from ground chicken meat. Lett Appl Microbiol 43:313–317.

Jasson V, Jacxsens L, Luning P, Rajkovic A, Uyttendaele M (2010) Review. Alternative microbial methods: An overview and selection criteria. Food Microbiol 27:710–730.

Kannan P, Yong HY, Reiman L, Cleaver C, Patel P, Bhagwat AA (2010) Bacteriophage-based rapid and sensitive detection of *Escherichia coli* O157:H7 isolates from ground beef. Foodborne Pathog Dis 7:1551–1558.

Kawasaki S, Fratamico PM, Horikoshi N, Okada Y, Takeshita K, Sameshima T, Kawamoto S (2009) Evaluation of a multiplex PCR system for simultaneous detection of *Salmonella* spp., *Listeria monocytogenes*, and *Escherichia coli* O157:H7 in foods and in food subjected to freezing. Foodborne Pathog Dis 6:81–89.

Leon-Velarde CG, Zosherafatein L, Odumeru JA (2009) Application of an automated immunomagnetic separation-enzyme immunoassay for the detection of *Salmonella enterica* subspecies *enterica* from poultry environmental swabs. J Microbiol Methods 79:13–17.

Lin A, Sultan O, Lau HK, Wong E, Hartman G, Lauzon CR (2011) O serogroup specific real time PCR assays for the detection and identification of nine clinically relevant non-O157 STECs. Food Microbiol 28:478–483.

López V, Suárez M, Chico-Calero I, Navas J, Martínez-Suárez JV (2006) Foodborne *Listeria monocytogenes*: are all the isolates equally virulent? Rev Argent Microbiol 38:224–234.

Manafi M (2000) New developments in chromogenic and fluorogenic culture media. Int J Food Microbiol 60:205–218.

Mandal PK, Biswas AK, Choi K, Pal UK (2011) Methods for rapid detection of foodborne pathogens: an overview. Am J Food Technol 6:87–102.

Mathusa EC, Chen Y, Enache E, Hontz L (2010) Non-O157 Shiga toxin-producing *Escherichia coli* in foods. J Food Prot 73:1721–1736.

Miller JM, Rhoden DL (1991) Preliminary evaluation of Biolog, a carbon source utilization method for bacterial identification. J Clin Microbiol 29:1143–1147.

Navas J, Ortiz S, Lopez P, Jantzen MM, Lopez V, Martinez-Suarez JV (2006) Evaluation of effects of primary and secondary enrichment for the detection of *Listeria monocytogenes* by real-time PCR in retail ground chicken meat. Foodborne Pathog Dis 3:347–354.

Neupane M, Abu-Ali GS, Mitra A, Lacher DW, Manning SD, Riordan JT (2011) Shiga toxin 2 overexpression in *Escherichia coli* O157:H7 strains associated with severe human disease. Microb Pathog 51:466–470.

O'Sullivan J, Bolton DJ, Duffy G, Baylis C, Tozzoli R, Wasteson Y, Lofdahl S (2007) Methods for detection and molecular characterisation of pathogenic *Escherichia coli*. Co-ordination action FOOD-CT-2006-036256. Pathogenic *Escherichia coli* network. Editors O'Sullivan J, Bolton DJ, Duffy G, Baylis C, Tozzoli R, Wasteson Y, Lofdahl S. http://www.antimicrobialresistance.dk/data/images/protocols/e%20coli%20methods.pdf. Accessed 15 November 2011.

Olsvik O, Popovic T, Skjerve E, Cudjoe KS, Hornes E, Ugelstad J, Uhlen M (1994) Magnetic separation techniques in diagnostic microbiology. Clin Microbiol Rev 7:43–54.

Pavlic M, Griffiths MW (2009) Principles, applications, and limitations of automated ribotyping as a rapid method in food safety. Foodborne Pathog Dis 6:1047–1055.

Pettipher GL, Watts YB, Langford SA, Kroll RG (1992) Preliminary evaluation of COBRA, an automated DEFT instrument, for the rapid enumeration of microorganisms in cultures, raw-milk, meat and fish. Lett Appl. Microbiol 14:206–209.

Posthuma-Trumpie GA, Korf J, van Amerongen A (2009) Lateral flow (immuno) assay: its strengths, weaknesses, opportunities and threats; a literature survey. Anal Bioanal Chem 393:569–582.

Pyle BH, Broadaway SC, McFeters GA (1999) Sensitive detection of *Escherichia coli* O157:H7 in food and water by immunomagnetic separation and solid-phase laser cytometry. Appl Environ Microbiol 65:1966–1972.

Rodriguez-Lazaro D, Hernandez M (2006) Isolation of *Listeria monocytogenes* DNA from meat products for quantitative detection by real-time PCR. J Rapid Method Automat Microbiol 14:395–404.

Samkutty PJ, Gough RH, Adkinson RW, McGrew P (2001) Rapid assessment of the bacteriological quality of raw milk using ATP bioluminescence. J Food Prot 64:208–212.

Savoye F, Feng P, Rozand C, Bouvier M, Gleizal A, Thevenot D (2011) Comparative evaluation of a phage protein ligand assay with real-time PCR and a reference method for the detection of *Escherichia coli* O157:H7 in raw ground beef and trimmings. J Food Prot 74:6–12.

Scheutz F, Nielsen EM, Frimodt-Møller J, Boisen N, Morabito S, Tozzoli R, Nataro JP, Caprioli A (2011) Characteristics of the enteroaggregative Shiga toxin/verotoxin-producing *Escherichia coli* O104:H4 strain causing the outbreak of haemolytic uraemic syndrome in Germany, May to June 2011. Euro Surveill 16 pii:19889.

Seo KH, Brackett RE, Frank JF, Hilliard S (1998) Immunomagnetic separation and flow cytometry for rapid detection of *Escherichia coli* O157:H7. J Food Prot 61:812–816.

Stager CE, Davis JR (1992) Automated systems for identification of microorganisms. Clin Microbiol Rev 5:302–327.

Stannard C (1997) Development and use of microbiological criteria for foods. Food Sci Technol Today 11:137–177.

Stevens KA, Jaykus LA (2004) Bacterial separation and concentration from complex sample matrices: A review. Crit Rev Microbiol 30:7–24.

Thacker JD, Casale ES, Tucker CM (1996) Immunoassays (ELISA) for rapid, quantitative analysis in the food-processing industry. J Agric Food Chem 44:2680–2685.

Torlak E, Akan IM, Gokmen M (2008) Comparison of TEMPO EC and TBX medium for the enumeration of *Escherichia coli* in cheese. Lett Appl Microbiol 47:566–570.

United States Department of Agriculture Food Safety And Inspection Service (2010a) Detection, isolation and identification of *Escherichia coli* O157:H7 from meat products, MLG 5.05, 10/01/10. http://www.fsis.usda.gov/PDF/MLG_5_05.pdf. Accessed 22 November 2011.

United States Department of Agriculture Food Safety And Inspection Service (2010b) FSIS procedure for the use of *Escherichia coli* O157:H7 screening tests, MLG 5A.02, 10/01/10. http://www.fsis.usda.gov/PDF/MLG_5A_02.pdf. Accessed 22 November 2011.

United States Department of Agriculture Food Safety And Inspection Service (2010c) Detection and isolation of non-O157 Shiga-toxin producing *Escherichia coli* strains (STEC) from meat products, MLG 5B.00, 10/01/10. http://www.fsis.usda.gov/PDF/MLG_5B_00.pdf. Accessed 7 November 2011.

Valadez AM, Debroy C, Dudley E, Cutter CN (2011) Multiplex PCR detection of Shiga toxin-producing *Escherichia coli* strains belonging to serogroups O157, O103, O91, O113, O145, O111, and O26 experimentally inoculated in beef carcass swabs, beef trim, and ground beef. J Food Prot 74:228–239.

van Griethuysen A, Bes M, Etienne J, Zbinden R, Kluytmans J (2001) International multicenter evaluation of latex agglutination tests for identification of *Staphylococcus aureus*. J Clin Microbiol 39:86–89.

Veal DA, Deere D, Ferrari B, Piper J, Attfield PV (2000) Fluorescence staining and flow cytometry for monitoring microbial cells. J Immunol Method 243:191–210.

Verstraete K, Robyn J, Del-Favero J, De Rijk P, Joris MA, Herman L, Heyndrickx M, De Zutter L, De Reu K (2012) Evaluation of a multiplex-PCR detection in combination with an isolation method for STEC O26, O103, O111, O145 and sorbitol fermenting O157 in food. Food Microbiol 29:49–55.

Wagner M, Horn M, Daims H (2003) Fluorescence in situ hybridisation for the identification and characterisation of prokaryotes. Curr Opin Microbiol 6:302–309.

Wawerla M, Stolle A, Schalch B, Eisgruber H (1999) Impedance microbiology: applications in food hygiene. J Food Prot 62:1488–1496.

Wu VCH (2008) A review of microbial injury and recovery methods in food. Food Microbiol 25:735–744.

Yang L, Bashir R (2008) Electrical/electrochemical impedance for rapid detection of foodborne pathogenic bacteria. Biotechnol Adv 26:135–150.

Chapter 3
DNA Microarrays: Principles and Technologies

Abstract A microarray is defined as a collection of microscopic features densely organized onto the surface of a solid support that can be probed with target molecules chemically labeled to produce either quantitative or qualitative data. For biological applications, the features that make up the array can be DNA, RNA, proteins, polysaccharides, lipids, small organic compounds, or even whole cells. DNA microarray technology is the most popular and well-developed usage of microarrays. It has allowed researchers to perform large-scale quantitative experiments and has contributed to fundamental changes in the way to perform biological research, moving from a gene-by-gene approach to global or genome-wide systemic studies.

Keywords DNA microarray technologies • Microarray methodologies • Microarray applications

3.1 Introduction

The idea of performing biological reactions with one spatially immobilized reagent is not new. In 1975, Edwin Southern described a technique that transformed molecular biology. Later during the 1980s, researchers began to work on a technique in which molecules of a known identity were immobilized on a membrane or slide and the solution to be tested was labeled and hybridized to the surface. Such arrays were used in DNA mapping and sequencing. This work concluded with a publication by Fodor et al. in 1991 about a new technology called Affymax (later Affymetrix). The paper described protein and nucleotide arrays, their production using photolithography, and their applications.

In 1995, Schena et al. introduced the word "microarray" for the first time. Initially, the technique was greatly limited, but the amount of sequencing data began to grow rapidly and the power of microarrays increased tremendously. By the mid-1990s, scientists had extended microarray capabilities by using them to study

G. López-Campos et al., *Microarray Detection and Characterization of Bacterial Foodborne Pathogens*, SpringerBriefs in Food, Health, and Nutrition, DOI 10.1007/978-1-4614-3250-0_3, © Guillermo Lopez-Campos, Joaquin V. Martinez-Suarez, Mónica Aguado-Urda, Victoria Lopez Alonso 2012

gene expression levels of hundreds of genes simultaneously, and in the process established microarrays as a viable and flexible molecular biology tool. The twenty-first century has witnessed an explosion in microarray-based publications, as researchers find more and more uses for the technology. It is difficult to find an issue of a contemporary biology journal that does not mention microarray technology (Wheelan et al. 2008).

3.2   DNA Microarray Principles and Technologies

The fundamental basis of DNA microarray technology involves the parallel hybridization of a mixture of nucleic acids (targets) with thousands of individual nucleic acids species (probes). Probes have a specific location on the array (spot or feature) so they can be identified by their spatial position in a single experiment. The probe sequences are immobilized in a tightly packed manner so that it is possible to place many different probes on a single small surface. While the probes are immobilized, the targets are deposited as a solution onto the array in order to hybridize. The target sample is usually labeled with a fluorescent dye that can be detected by a light scanner that scans the surface of the microarray. The key to microarray technology is that a probe is detected at a level proportional to the amount of its target present in the labeled extract.

DNA microarrays have been widely used for gene expression profiling and genotype analysis. Observing all the microarray spots at the same time gives the profile of a sample; in the most common application of DNA microarrays, the target sample is mRNA and the total microarray image represents the transcriptional profile of the sample (de Rinaldis and Lahm 2006). Although mRNA expression profiling is the dominant application for DNA microarrays, they are also used in genomic DNA analyses such as the detection of alternatively spliced variants, the epigenetic status of the genome, DNA copy number changes and sequence polymorphisms, or to detect DNA–protein interactions. The potential applications include the analysis and characterization of any reaction product composed of nucleotide sequences (Shiu and Borevitz 2008).

Due to advances in manufacture, robotics, and bioinformatics, microarray technology has continued to improve in terms of efficiency, reproducibility, sensitivity, and specificity. These improvements have allowed microarrays to transition from strictly the research setting to clinical diagnostic applications (Miller and Tang 2009). In addition to the ability to examine a large number of genes in parallel, the success of microarrays can be attributed to the versatility and flexibility of array designs. Currently, many microarray platforms are available, and custom array designs are possible and relatively cost-efficient. DNA microarray technology can be classified according to the method used for the deposition of the probe sequences onto the solid surface. This can be done by presynthesis (printed microarrays) or in situ synthesis (in situ-synthesized oligonucleotide microarrays). The solid surface used varies depending on the type of microarray employed.

There are downsides to microarray technology. Microarrays are still expensive, and more importantly, microarray experiments generate enormous amounts of complex data that are not analyzed easily. There is currently no consensus about how many repetitions need to be done, although the number fortunately appears to be low. The cost of arrays is independent of the cost of the reagents (Cy3 and Cy5 are expensive). The array scanner and workstation, the time and effort of a technician and an analyst, and of course the cost associated with the time spent doing the experiment in the first place must all be added to the cost of microarray experimentation (Wheelan et al. 2008).

3.2.1 Printed DNA Microarrays

Arrays based on presynthesis of the probes are known as spotted or printed arrays because the probes are "printed" or spotted onto the microarray surface. In this technology, a robot spotter is used to place small quantities of a probe in solution on the microarray surface (commonly a glass microscope slide). Glass slides are an attractive medium for microarrays because they are economical, stable during high temperatures and stringent washes, nonporous (allowing for efficient kinetics during hybridization), and have minimal background fluorescence (de Rinaldis and Lahm 2006).

The probe spots can be applied by either noncontact or contact printing. A noncontact printer shoots small droplets of probe solution onto the glass slide. In contact printing, each print pin directly applies the probe solution onto the microarray surface. The result is the application of probe solution to create spots of 100–150 μm. Control for cross-contamination and consistency during the printing process is crucial to preserve the reliability of the microarray and consequent hybridization data. Due to the relatively large size of the features, printed microarrays are of lower density (10,000–30,000 spots) than in situ-synthesized microarrays or high-density bead arrays, but offer considerably more features than suspension bead arrays (Miller and Tang 2009). The main advantage of the presynthesis of oligos is the ability to generate arrays with specifically desired sequences. The presynthesis technology is often used in small laboratories that want to have the freedom to target specific sequences, which are often not standard, and therefore need to design their own chips. This flexibility allows even the spotting of unknown DNA sequences, which can be essential for particular experimental designs.

According to the nature of the probe, printed arrays can be classified as double-stranded DNA microarrays or oligonucleotide microarrays. For double-stranded (ds) DNA microarrays, the probes consist of amplification products (amplicons) obtained by PCR using primers designed from a known genomic sequence, shotgun library clones, or cDNA. The double-stranded amplicons are denatured to allow the probes to be available for hybridization. Amplicons can be attached to the glass slide surface either by the electrostatic interaction between the negative charge of the phosphate backbone of the DNA and a positively charged coating on the slide surface or by UV-cross-linked covalent bonds between the thymidine bases in the DNA and amine groups on treated slides. A 200–800-bp dsDNA probe is recommended, but larger fragments of up to 1.3 kb in length also work. In a typical microarray design, each probe corresponds to one gene. PCR amplicons for microarrays should produce a high yield and specificity and no contamination (nonspecific amplification or contaminants that autofluoresce themselves or affect the attachment of targets to the microarray surface). Nevertheless, the generation of whole-genome DNA microarrays by high-throughput PCR amplification is a very laborious process, and dsDNA probes generally have high sensitivity but suffer in specificity. Extensive quality control by gel electrophoresis, product purification, and repetition of dropout reactions is necessary (Ehrenreich 2006). The ultimate assessment of probe specificity requires sequencing the products. Decreased specificity can be beneficial when analyzing a genomic sequence rich in natural polymorphisms, but it is disadvantageous when trying to discriminate between highly similar target sequences. Moreover, high specificity is necessary for clinical diagnostic applications (Miller and Tang 2009). Two advantages that spotted dsDNA microarrays have are their higher hybridization sensitivity and their lower cost.

Oligonucleotide spotted microarray probes consists of short chemically synthesized sequences. Usually, the length of the probes ranges from 25 to 80 bp but may be as long as 150 bp for gene expression microarrays. Using oligonucleotides as probes is an alternative to double-stranded DNA because fewer errors are introduced during probe synthesis and it allows for the testing of small genomic regions, including polymorphisms. A decreased probe length may adversely affect the sensitivity compared to dsDNA probes, but the specificity is greater when short specific genomic regions are interrogated. Moreover, longer probes have higher melting temperatures and greater mismatch tolerance, leading to decreased specificity. Generally, the larger the length of the probe, the stronger the hybridization signal and the higher the sensitivity. Because of the small size of the probes in general, oligonucleotides are commonly attached to the slide surface by covalent coupling. Otherwise, a significant amount of probes would be lost during hybridization and washing. To achieve covalent linkage, slides are commonly coated with compounds with aldehyde or epoxy functional groups. Oligonucleotides are attached to the microarray surface via a modified 5' or 3' end. This increases the availability of the probe sequences for hybridization with potential targets, because they are not fixed to the surface by the nucleic acid backbone or by individual bases. A further improvement in sensitivity can be achieved by inserting spacer molecules between the oligonucleotides and the slide surface (most commonly a 5' amino group) (Ehrenreich 2006). Despite being easier to manufacture than dsDNA probes, oligonucleotide probes need to be designed carefully so that all the probes have similar melting temperatures (within a range of 5°C) and lack palindromic sequences.

Printed microarrays are relatively simple and inexpensive compared to in situ-synthesized microarrays. Setting up microarray facilities is costly and requires a dedicated space in which the environment is well controlled (dust, humidity, temperature, ozone levels, etc.). However, if microarray core facilities are available at research centers and/or universities, these challenges don't affect the individual researchers as much. The huge scale of amplicon production along with the issues of quality control, information management, efficiency, and accuracy are the main difficulties in the manufacturing of printed dsDNA microarrays. Moreover, the design of oligonucleotide probes is laborious, and errors introduced during probe synthesis are a problem. There are a great number of commercially available whole-genome printed microarrays for select organisms. Additionally, printed dsDNA microarrays are important for the study of organisms that have not been fully sequenced.

3.2.2 In Situ-Synthesized Oligonucleotide Microarrays

For microarrays synthesized in situ, the manufacturing steps of cloning, amplification by PCR, and probe spotting are not necessary. This offers an important advantage by reducing the noise and variability of the system. Most array manufacturers offer standard ready-to-use chips designed for monitoring the expression profiles or the genotype of many common organisms such as human, mouse, rat, or yeast. In this case, the customer is not responsible for the complex problems of probe design or attaching the probes to the chip.

In situ-synthesized arrays are high-density microarrays that, like oligonucleotide printed arrays, use oligonucleotide probes. In contrast to printed oligonucleotide arrays, however, the oligonucleotide probes are synthesized directly on the surface of the microarray.

Arrays generated by this approach achieve DNA oligonucleotide synthesis by successive rounds of deprotection from UV light with many photolithographic masks. By combining solid-phase chemistry and photolithography, extremely high-density oligonucleotide arrays containing more than 10^6 features can be manufactured. GeneChips (Affymetrix) are the most commonly known. In situ-synthesized probes are usually short (20–25 bp), and typically, multiple probes per target are included to improve the sensitivity, specificity, and statistical accuracy. The use of probe sets is common to increase the specificity. A probe set includes one perfect match probe and one mismatch probe that contains a 1-bp difference in the middle position of the probe. Results from the exact probe versus the mismatched probe can therefore be used to detect and eliminate cross-hybridization.

GeneChips oligonucleotide probes (Affymetrix) are synthesized using semiconductor-based photochemical synthesis. There are synthetic linkers on the quartz surface modified with light-sensitive protecting groups. Consequently, the microarray surface is chemically protected from a nucleotide attachment until deprotected by light. Once the array surface is exposed to UV light, reactive nucleotides modified with a photolabile protecting group can be added to growing oligonucleotide chains. Photolithographic masks are used to target specific nucleotides to exact probe sites. Each photolithographic mask has a defined pattern of windows, which act as a filter to either transmit or block UV light. Areas of the microarray surface in which UV light has been blocked will remain protected from the addition of nucleotides, whereas areas exposed to light will be deprotected, and specific nucleotides can be added. The pattern of windows in each mask directs the order of nucleotide addition. In situ probe synthesis is therefore achieved through the repeated cycling of masking, light exposure, and the addition of either A, C, T, or G bases to the growing oligonucleotides (Miller and Tang 2009).

Roche NimbleGen and Agilent Technologies are other important manufactures of high-density oligonucleotide microarrays. In both platforms, the oligonucleotide probes used are longer (60–100 bp). NimbleGen microarrays can contain more than 10^6 features, and different formats per slide are available, such as 1×2.1 million features, $3 \times 720,000$ features, $1 \times 385,000$ features, $4 \times 72,000$ features, and $12 \times 135,000$ features. Agilent microarrays are available in the following formats: SurePrint G3 formats ($8 \times 60,000$ features, $4 \times 180,000$ features, $2 \times 400,000$ features, and $1 \times 1,000,000$ features), and SurePrint HD formats ($8 \times 15,000$ features, $4 \times 44,000$ features, $2 \times 105,000$ features, and $1 \times 244,000$ features).

While NimbleGen manufactures probes by maskless photo-mediated synthesis similar to the process used for Genechips, Agilent employs inkjet technology for in situ synthesis. Affymetrix and NimbleGen microarray technologies use quartz wafers, while Agilent microarrays use glass slides and inkjet printing, which eliminates the need for either lithographic or digital masks. Unlike Affymetrix chips, which are hybridized with only one labeled target, the NimbleGen and Agilent platforms allow multicolor hybridizations and the use of longer oligonucleotides, thus increasing sensitivity.

Synthesized microarrays depend on commercial manufacturing because of the complexity behind chemical synthesis and the high cost involved in their production. The number of microbial genome microarrays that are commercially available for gene expression studies is continually growing. For specific applications, synthesized oligonucleotide arrays can be designed and ordered as custom microarrays. A custom Affymetrix microarray can be quite expensive, and the inflexibility of its custom mask makes the use of an Affymetrix-synthesized array impractical for some applications. In contrast, the NimbleGen and Agilent platforms are easily customized with a unique oligonucleotide sequence content. In addition, a web-based tool provided by Agilent called eArray allows users to design custom microarrays with no minimum manufacturing batch size requirements, making Agilent microarrays a primary choice for homebrew and pilot applications. The major advantages to these systems are the reproducibility of the manufacturing process and the standardization of reagents, instrumentation, and data analysis. Oligonucleotide microarrays generally allow much cleaner downstream hybridization data than amplicon-based microarrays. With oligonucleotide arrays, the ability to standardize probe concentrations and hybridization temperatures has resulted in considerable improvements in the accuracy and reproducibility of microarray data (Miller and Tang 2009).

3.2.3 Suspension Bead Arrays

Suspension bead arrays are three-dimensional arrays based on the use of microscopic polystyrene spheres as a solid support and flow cytometry for bead and target detection. Multiplexing is accomplished by using different microsphere sets based on color (Miller and Tang 2009).

The current generation of commercially available suspension bead arrays (LabMap System, Luminex) offers a 100-element array that uses a flow cytometer for analysis of the microspheres in suspension. The flow cytometer provides a method to simultaneously detect DNA-binding events on each microsphere. Two spectrally distinct fluorophores are incorporated on polystyrene microspheres (FlowMetrix Beads). Using defined ratios of these fluorophores, a suspension microarray is created consisting of up to 100 different microsphere sets. The flow cytometer provides two lasers for simultaneous excitation of both fluorophores. A third fluorophore coupled to the reporter molecule quantifies the biomolecular interaction that has occurred on the surface of each microsphere. A suspension of microspheres is individualized by the hydrodynamic focusing effect of a flow cell. High-speed digital signal processing classifies the microspheres according to their spectral properties and thus is able to quantify the reaction on the surface. For each analyte, a defined quantity of microspheres is added to the sample. After mixing and incubating the analytes and microspheres, the detector molecules are added. A centrifugation or filtration step is needed after incubation to separate the unbound components. The washed bead suspension is directly read with the flow cytometer (Seidel and Niessner 2008).

Although the feature density of suspension bead arrays is the lowest of the platforms that have been reviewed here, the relative simplicity, powerful multiplexing capabilities, and relatively low cost make this platform the most practical system for high-throughput nucleic acid detection in applications such as clinical diagnosis of infectious diseases (Miller and Tang 2009).

3.2.4 High-Density Bead Arrays

BeadArrays (Illumina) offer an attractive substrate for the high-density detection of target nucleic acids. BeadArrays are based on 3-μm silica beads that randomly self-assemble onto one of two available substrates: the Sentrix Array Matrix (SAM) or the Sentrix BeadChip. The SAM contains 96 1.4-mm fiber-optic bundles. Each bundle is an individual array consisting of 50,000 5-μm light-conducting fibers, each of which is chemically fixed to create a microwell for a single bead. Each SAM allows the analysis of 96 independent samples. The Sentrix BeadChip is used to assay 1–16 samples at the same time on a silicon slide that has been processed by microelectromechanical systems technology to provide microwells for individual beads. BeadChips are more appropriate for very high-density applications such as whole-genome genotyping and for determining genome-wide single-nucleotide polymorphisms. In BeadArrays technology, the beads randomly separate to their final location on the array so the bead location must be mapped, which is in contrast to the known locations of printed or in situ-synthesized microarrays. The mapping is accomplished by a decoding process. Each bead has approximately 700,000 copies of a unique capture oligonucleotide covalently attached to it, which acts as the bead's identifier. After a series of hybridization and rinse steps, fluorescently labeled complementary oligonucleotides bind to their specific bead sequence, making the identification of the bead location possible (Seidel and Niessner 2008; Miller and Tang 2009). This decoding process offers the additional advantage of the quality control provided for each feature of the microarray. Since each manufactured microarray will not be identical, BeadArrays have built-in redundancy, which provides a crucial experimental control for inter-microarray comparative data (Miller and Tang 2009). BeadArrays have been successfully used for DNA methylation studies, gene expression profiling, and SNP genotyping.

3.3 Microarray Approaches and Applications

Regardless of the technology used, the final aim of any microarray assay is to provide a measure for each probe of the relative abundance of the complementary target in the sample of interest. When well designed, microarrays can answer a huge number of questions with a single experiment or can increase the scope of an experiment to previously inaccessible levels (Wheelan et al. 2008). Thus, microarrays have been successfully used in a variety of applications, including sequencing, single-nucleotide polymorphism (SNP) detection, characterization of protein–DNA interactions, DNA computing, and others. However, mRNA profiling applications currently dominate microarray usage because of the amount of information that can be obtained about the functions of genes in cells and tissues. Applications can be divided into two main classes: mRNA gene expression profiling and analysis of genomic DNA.

3.3.1 mRNA Gene Expression Profiling

Undoubtedly, the most often used application of DNA microarray technology is transcription analysis. The transcriptome, or the entire set of genes transcribed in a specific cell at a given time under defined conditions, can effectively be analyzed using DNA microarrays. However, it is essential to spend some time thinking about exactly what type of gene expression changes are expected and in what type of cells those changes would be best detected. Microarray experiments are relatively easy to perform, but poor experimental design may yield results that are difficult or impossible to interpret (Ness 2006).

- *Case versus control studies.* The most common and basic question in DNA microarray experiments is whether genes appear to be upregulated or downregulated between two or more groups of samples. This type of analysis is essential because it provides the simplest characterization of the specific molecular differences associated with a specific biological effect. These signatures can be used to generate new hypotheses and guide the design of further experiments (Trevino et al. 2007). Treating a cell line or microorganism with a specific treatment condition generates immediate and rapid changes in gene expression that can be detected with microarray assays (Ness 2006). When comparing two biological conditions such as disease state vs. normal state, genes that are differentially expressed in the disease state can be identified and hypotheses can be made to identify the genes that play a causal role in the development of the disease. If the transcriptional data are also confirmed at the protein level, these genes can potentially pose an interest as candidates for drug targets in pharmaceutical research. Ideally, drugs could be designed to specifically inhibit any particular gene, protein, or signaling cascade, and if the target is specifically expressed in the diseased tissue, there is less chance of causing undesirable effects (de Rinaldis 2006). Simple case vs. control studies have given way to more powerful experimental designs to suggest targets and illuminate disease mechanisms.

- *Comparison of samples.* Microarray technology offers a rapid and sensitive way to compare gene expression profiles in tumors from different individuals as a potential clinical tool to identify which types of tumors might respond better to a particular treatment or for identifying patients with better or worse prognoses. Such information could be particularly helpful to make decisions about which therapeutic options are most appropriate. Nevertheless, these studies involve quite complex data and statistical analyses. Additionally, successful clinical studies require balanced cohorts designed by qualified biostatisticians to avoid common pitfalls and artifacts (Ness 2006).
- *Functional response patterns.* The power of expression profiling is most evident in experiments that explore a systematically varied set of conditions. Data redundancy is provided by sampling a smoothly varying process. Co-regulation of genes across a set of biological conditions can reveal functional gene groups (Stoughton 2005). These co-regulation–based groupings are a fairly accepted way to gain functional information using the "guilt by association" inference. In other words, a gene of unknown function is predicted to be associated with a functional role if its expression pattern is similar to that of a gene of known function (Quackenbush 2003).
- *Studying pathways and biological gene networks.* While standard gene expression analysis looks at each gene as an independent entity, pathways analysis is designed for the identification of coordinated changes in expression, affecting many genes at the same time. The idea at the basis of the various approaches and tools for pathway analysis is that of characterizing the behavior of groups of genes that act in concert to carry out a specific function. Defined gene sets can include annotation-based groups of genes belonging to the same functional category/pathway or can derive from the analysis of independent expression data, like clusters of co-expressed genes, genes expressed in a particular tissue type or in particular conditions (de Rinaldis and Lahm 2006).

3.3.2 Analysis of Genomic DNA

Genomes constantly experience expansion, contraction, and other changes due to events such as deletion, insertion, translocation, and inversion. These changes can be identified through comparative genomic hybridization (CGH), where DNA sequence differences throughout the entire genome are monitored by comparing differentially labeled test DNA and reference DNA (Shiu and Borevitz 2008).

- *Genotyping.* The rapid acquisition of genetic information was one of the original objectives of the development of Affymetrix microarray technology. In 1996, Chee et al. demonstrated a method of resequencing for point mutations using microarrays, which currently has become an established technique. The baseline method involves short probes complementary to every N-mer of the baseline target sequence along with additional probes with each of the four nucleotides at the putative mutation position. Each of these also can be paired with a "mismatch" probe to control for nonspecific hybridization. Chips have been designed for mutation detection in genes of particular interest to human health, showing the promise of these methods and also the difficulties associated with false detections when the underlying mutation rates are low (Stoughton 2005).
- *Detailed characterization of microbial pathogens.* By enabling parallel interrogation of pathogen genomes, microarrays offer promising improvements in the diagnosis of infectious diseases, monitoring of emerging infections, and examining the safety of food, water, and air (Stoughton 2005). Probes targeted to specific genes are used to detect the presence of virulence factors, antigenic determinants, and drug resistance determinants as well as to closely resolve related species of bacteria. Probes designed for genes of a baseline strain can be used to characterize and compare the genomes of test strains via competitive hybridization. CGH can be conducted both within species and between species to discover instances of gene gain or loss. CGH is not restricted to genes, since arrays with nongenic sequences or even whole genomes are commercially available for several model organisms or can be fabricated on demand (Shiu and Borevitz 2008). Host–microbe interactions also could be studied in detail using a combination of genomic analysis of the pathogen and expression profiling of host immune cells (Stoughton 2005).

3.4 Microarray Methodologies

Microarray technology allows simultaneous massive parallel determination and multiple measurements of a variety of binding events. Moreover, it has the advantage of requiring a small amount of material, and it may easily be automated with the possibility of saving a great deal of time.

In a typical expression microarray experiment, the system relies on measuring the absolute intensity of the labeled target from each sample. In a one-color strategy, mRNA from one type of cells or tissues is isolated and reverse-transcribed to cDNA to avoid degradation. Usually, fluorescently labeled nucleotides are incorporated during the reverse transcription. After the labeling, cDNAs are hybridized on the microarray, which is then analyzed by a laser scanner, and the intensity of each feature is measured.

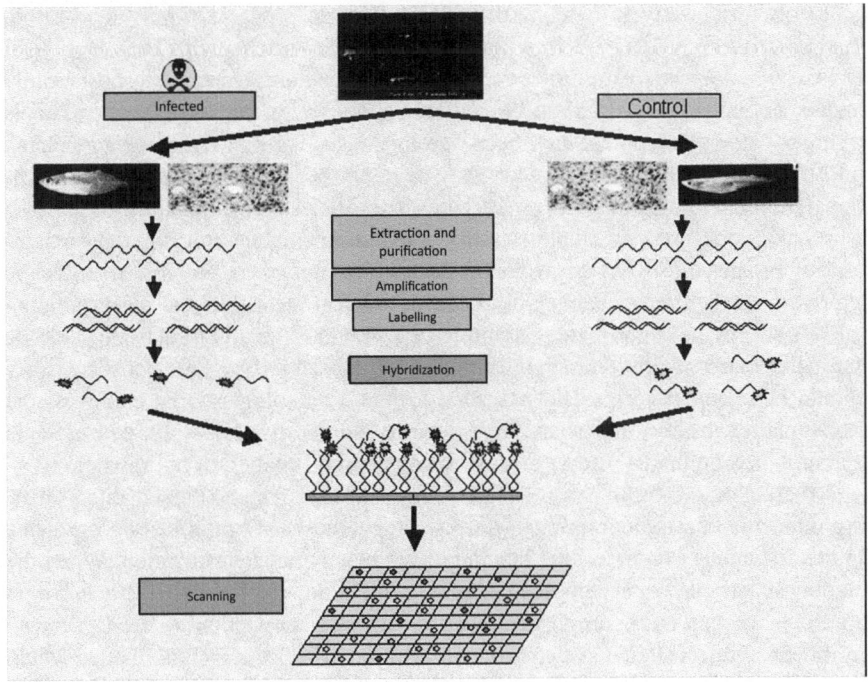

Fig. 3.1 Steps of a two-color approach strategy to do a comparative genomic hybridization experiment with DNA microarrays

A different commonly used technique to measure gene expression involves RNA isolation from two separate samples. This is the two-color approach strategy, a comparative hybridization experiment between two samples. In this technique, the two different samples are each labeled with two different dyes that fluoresce at nonoverlapping wavelengths (typically, Cy3 and Cy5 are used). The two pools of labeled cDNAs are mixed and then hybridized on the array. If expression levels are similar for both samples, then both Cy3- and Cy5-labeled cDNA are present, resulting in a yellow-colored spot. If expression levels from the Cy5-labeled cDNA are higher, then a red spot will appear. On the contrary, if expression levels from the Cy3-labeled cDNA are higher, then a green spot will be observed (Fig. 3.1).

3.4.1   Sample Preparation

There are many variations in the protocol for isolating and amplifying target nucleic acid. In the case of mRNA profiling, either mRNA or total RNA can be the starting material for amplification. In prokaryotes, mRNA purification is problematic because the nucleic acid lacks 3′ polyadenylation, and there are no widely adopted protocols for selectively labeling the mRNA. Due to the lack of polyadenylation, random priming has to be used with either hexamers or nonamers. Thus, only total RNA can be labeled, resulting in a higher background and requiring a substantially higher amount of total RNA to be added to the labeling reaction. New protocols trying to solve this problem have been developed, such as preparing polyadenylated mRNA from prokaryotes; however, they have not been widely adopted. An additional problem in working with prokaryotic mRNA is its higher instability compared with eukaryotic mRNA. This instability demands special care during the preparation of prokaryotic RNA to avoid producing artifacts. It is possible to miss the expression of certain genes simply because of degradation of the corresponding mRNA during the experiment (Ehrenreich 2006). Good-quality total bacterial RNA can be prepared with commercial kits such as RNeasy (Qiagen), but the quality should be checked with a Bioanalyzer (Agilent Technologies) that expresses the RNA quality in the form of an RNA integrity number (RIN). RNA with a value higher than 8 on the RIN scale of 0–10 is considered suitable for hybridization.

Labeling molecules or modified nucleotides capable of being tagged with a labeling molecule can be incorporated during the synthesis of amplification products. In the Affymetrix protocol, the labeling takes place after hybridization. When the amount of sample material is scarce, an amplification step is needed. Nucleic acid amplification can be accomplished through reverse transcription of RNA followed by linear amplification via in vitro transcription (IVT) and/or polymerase chain reaction (PCR). Amplification can be 3'-biased or full-length, and this decision influences the process of probe design. 3'-Biased amplification methods take advantage of the ability to prime from polyadenylation sequences found in eukaryotic transcripts. Full-length amplification tends to employ random priming of the target molecules, either because poly(A) sequences do not exist (prokaryotic organisms) or because of a desire to obtain amplification product lengths that are more representative of the complete target sequences. The final product to be hybridized to the array can be either cDNA or cRNA. Linear and modest amplification, as well as postsynthetic incorporation of labels, are usually associated with more reproducible data. cDNA–DNA hybridizations are likely to suffer less from cross-hybridization, although the binding energies tend to be lower than those of cRNA–DNA duplexes (Stoughton 2005).

3.4.2   Hybridization

During hybridization, complementary sequences should be able to find each other. The fundamental parameters are time, stringency, concentration, and complexity of the sample and density of available binding sites. Other factors include the distribution of fragment lengths, steric effects of dye molecules, and surface chemistry. The optimization of stringency involves choosing conditions in which, for most probes, perfect match duplexes have a high occupancy compared to mismatch duplexes. Overall, any given target sequence will pair and dissociate many times during the hybridization reaction, staying longer at high-binding-energy well-matched duplexes than in poorly matched duplexes. During this annealing, specificity increases over time as the reaction approaches equilibrium. In fact, the progression with time can be used to distinguish specific from nonspecific binding. Generally, long hybridization times at a relatively high stringency are required to obtain the best specificity (Stoughton 2005).

The hybridization of DNA microarrays can be done in two different ways. The classical approach includes placing the labeled target on a slide and carefully covering it. This requires some degree of skill to prevent gradients in hybridization and to avoid air bubbles. A hybridization chamber is usually needed to keep the temperature and humidity constant. Hybridization temperatures range from 40°C to 65°C depending on several factors, including the organism studied, the hybridization buffer composition, and the technology used. The hybridization temperature is critical for oligonucleotide microarrays and has to be carefully optimized. Hybridization solutions contain saline sodium citrate (SSC), sodium dodecyl sulfate (SDS) as detergent, nonspecific DNA such as salmon sperm DNA, blocking reagents to reduce the background like bovine serum albumin (BSA) or Denhardt's reagent, and labeled cDNA from the samples. Other factors such as agitation, microfluidic circulation (Affymetrix system), sonication, and the use of surfactants or buffers have the potential to speed up and improve hybridization. Washing off the unbound sample after hybridization is a crucial step. Stringency must be optimized here as well. More stringent washing steps are performed at the end of the washing procedure, which can be achieved by either decreasing the ionic strength or increasing the washing temperature (Ehrenreich 2006). Finally, the slides are dried and scanned within several hours after hybridization, because the fluorescence signal deteriorates with time. It is important to protect the slides from high levels of atmospheric ozone due to the sensitivity of some dyes (particularly during the drying step when the array surface is exposed to the air).

Automatic array hybridization stations are an alternative to the classical approach. The hybridization and washing steps are achieved by running programmed protocols. The results do not depend on the skill of the researcher and are very reproducible. Since the hybridization and washing conditions have to be fine-tuned according to the chemistry of the slide, the conditions are most easily optimized when a large number of arrays based on the same chemistry have to be handled identically. This is the case of the Affymetrix system.

3.4.3 Image Capture

After hybridization, the slide is read by a scanner, which consists of a device similar to a fluorescence microscope coupled with a laser, robotics, and a digital camera to record the fluorescent emission. Fluorescent labeling and detection on microarrays have replaced radioactive labeling because of the higher sensitivity and the fact that nonradioactive labeling is much easier and safer to handle. The amount of signal detected is presumed to be proportional to the amount of dye at each feature in the microarray and hence proportional to the target nucleic acid concentration of the complementary sequence in the sample. The output is a monochromatic digital image file typically in TIFF format for each fluorescent dye. False-color images containing red, green, and yellow are reconstructed by specialized software for visualization purposes only (Trevino et al. 2007).

Scanning of the fluorescent hybridization signal can be done with CCD imaging using filtered white light illumination. However, nowadays, it is common to use laser confocal scanners. The laser confocal approach has fundamental advantages, such as better signal-to-background ratios and less photobleaching of the labels (Stoughton 2005). In addition to scanners specific for Affymetrix technology, leading manufacturers of scanners include Agilent Technologies, Axon Instruments, and Genomic Solutions. Most devices have lasers and filter sets that are compatible with common fluorescent label pairs such as Cy3 and Cy5. Usually, scanners possess their own image processing software to reduce the raw images to spot intensities.

New labeling options, such as quantum dots and plasmon resonance particles, may finally allow efficiency as good as single-molecule detection and reduce requirements in amplification and quantity of biological sample input. Plasmon surface resonance (PSR) detection of molecular binding is potentially a way to scan microarrays without using any label, although this has yet to produce any commercial systems. PSR is an optical technique that investigates what happens at the interface of a thin metal-coated prism in contact with a solution, which is used to determine refractive index changes at the surface. When light is incident on the prism side at a particular angle (resonance angle), the intensity of the reflected light is at its minimum. In the presence of biomolecules on the metal surface, this angle variation is very sensitive. Changes in reflectivity give a signal that is proportional to the mass of the biomolecules bound to the surface. To detect one molecule, such as a DNA target, the ligand (probe) is immobilized onto the surface. As the target binds to the ligand, the mass and the refractive index increase; thus, detection of binding can be achieved without any label (Sassolas et al. 2008).

References

Chee M, Yang R, Hubbell E, Berno A, Huang XC, Stern D, Winkler J, Lockhart DJ, Morris MS, Fodor SP (1996) Accessing genetic information with high-density DNA arrays. Science 274:610–614.

de Rinaldis E, Lahm A (2006) DNA microarrays: Current Applications. Horizon Bioscience, Norfolk, UK.

Ehrenreich A (2006) DNA microarray technology for the microbiologist: an overview. Appl Microbiol Biotechnol 73:255–273.

Fodor SP, Read JL, Pirrung MC, Stryer L, Lu AT, Solas D (1991) Light-directed, spatially addressable parallel chemical synthesis. Science 251:767–773.

Miller MB, Tang YW (2009) Basic concepts of microarrays and potential applications in clinical microbiology. Clin Microbiol Rev 22:611–633.

Ness SA (2006) Basic Microarray Analysis. In: Larson RS (ed) Methods in Molecular Biology, vol. 316: Bioinformatics and Drug Discovery. Humana Press Inc. Totowa, New Jersey.

Quackenbush J (2003) Genomics. Microarrays – guilt by association. Science 302:240–241.

Sassolas A, Leca-Bouvier BD, Blum LJ (2008) DNA Biosensors and Microarrays. Chem Rev 108:109–139.

Schena M, Shalon D, Davis RW, Brown PO (1995) Quantitative monitoring of gene expression patterns with a complementary DNA microarray. Science 270:467–470.

Seidel M, Niessner R (2008) Automated analytical microarrays: a critical review. Anal Bioanal Chem 391:1521–1544.

Shiu SH, Borevitz JO (2008) The next generation of microarray research: applications in evolutionary and ecological genomics. Heredity 100:141–149.

Southern EM (1975) Detection of specific sequences among DNA fragments separated by gel electrophoresis. J Mol Biol 98:503–517.

Stoughton RB (2005) Applications of DNA microarrays in Biology. Ann Rev Biochem 74:53–82.

Trevino V, Falciani F, Barrera-Saldaña HA (2007) DNA Microarrays: a Powerful Genomic Tool for Biomedical and Clinical Research. Mol Med 13:527–541.

Wheelan SJ, Murillo FM, Boeke JD (2008) The incredible shrinking world of DNA microarrays. Mol Biosyst 4:726–732.

Chapter 4
Bioinformatics in Support of Microarray Experiments

Abstract For a long time, microarrays and bioinformatics have been intricately linked, and microarrays have emerged as a hot topic in the field of bioinformatics. This happened due to the particularities of microarray technology, where huge amounts of information both are needed for the design of experiments and are generated from such experiments. Therefore, microarrays cannot be properly used or understood without considering all the bioinformatical aspects associated with them. This chapter is dedicated to describing the relevance and importance of bioinformatics in the development of microarray experiments. Microarray technology has to deal with massive amounts of information gathered from different sources during experimentation, such as samples, devices, methodologies, and eventually, experimental results. Therefore, microarray data management and analysis still remain among the most challenging aspects of microarray-based approaches and experiments.

Keywords Microarray data analysis • Bioinformatics • Gene expression • Microarray databases • Microarray probe design

4.1 Introduction

Bioinformatics provides a great deal of support to microarray methodologies by providing tools and methods for the exploitation of the huge amounts of data and information generated during these experiments. The final outcome of a microarray experiment relies quite often on the correct use of all these bioinformatics tools. This chapter will be divided into three major topics dealing with the most relevant aspects related to the use of bioinformatics for successful microarray development. These major topics are

- Microarray probe design. One of the major challenges in developing a microarray experiment is deciding which probe should be attached to the surface of the microarray and which detection capabilities the system will use.

G. López-Campos et al., *Microarray Detection and Characterization of Bacterial Foodborne Pathogens*, SpringerBriefs in Food, Health, and Nutrition, DOI 10.1007/978-1-4614-3250-0_4,
© Guillermo Lopez-Campos, Joaquin V. Martinez-Suarez, Mónica Aguado-Urda, Victoria Lopez Alonso 2012

- Microarray data analysis. This topic covers the analysis steps involved from image capture to generation of the final results.
- Microarray data management. Due to the huge amounts of information generated during microarray experiments, it is of vital importance to manage and store this information correctly.

Selection of the different bioinformatics tools and strategies needed in a particular project for the analysis of foodborne-related pathogens is tightly linked to the experimental approach chosen for the experiment. In many cases, the selection of a particular approach or microarray technology limits or forces the adoption of different tools.

4.2 Microarray Probe Design

"Probes" refer to those molecules attached to the surface of the microarray. These probes are responsible for the detection of target molecules present in the samples, and therefore, an incorrect design may prevent the detection of a particular target or even lead to a more complicated situation by providing a misleading result. Therefore, microarray probe design is a key step in the development of any microarray-based experiment. The process of designing or choosing probes for the experiments is strongly dependent on the microarray platform and the experimental design chosen for the experiment. Different samples and targets require different probes. Similarly, requirements for different experimental approaches for the detection of a given target require different probe design characteristics. This chapter will focus on the design of microarray probes for DNA microarrays, since this type of microarray is the one most commonly used for the detection of foodborne pathogens.

The design of probes for DNA microarrays should be considered a challenge in bioinformatics as it deals with sequence analysis. Thus, the use of bioinformatics tools for sequence analysis is essential for this process. In chapter 3, the different experimental approaches available in microarray-based experiments have been described; each of these experimental approaches has its own specific requirements and parameters for probe design. A flow diagram with the major characteristics associated with the design of microarray probes, including some links to online tools, can be found in Biotic (2011).

Microarray technology relies on previously available data for probe design, as the detection of any biomarkers of interest requires available sequence information. The first step in probe design is defining the biomarkers of interest that will be analyzed on the microarray and analyzing the available sequences required for effective probe design. Prior to the point where probes are properly designed and their characteristics are chosen, it is necessary to retrieve the target sequences.

Sequence retrieval is a very important process since it is here when the user provides the required sequences that will be processed and analyzed to eventually become the immobilized probes on the microarray surface. This a priori needed sequence information and sequence data might be obtained from private databases, private sequencing projects, or public sequence databases. Public sequence databases can be categorized into two major types:

- Major general sequence databases. In these databases, it is possible to find all kinds of sequences from any organism. The three main sequence databases are Genbank Nucleotide at the National Center for Biotechnology Information (NCBI) site (Benson et al. 2011), EMBL-Bank at the European Bioinformatics Institute (EBI) (Cochrane et al. 2009), and DNA Data Bank of Japan at the Center for Information Biology and DNA Databank (Kaminuma et al. 2011). These three databases share DNA sequence data and are a general data source for sequences for probe design. Sequences available in these databases are not curated; therefore, their quality and any accompanying annotations rely simply on the information submitted by the users who uploaded them.
- Oriented databases. These resources are focused on specific organisms or sequences and generally contain sequences of higher quality and with better annotations. In many cases, the use of these resources simplifies the search and retrieval of particular sequences of interest for probe design. Some examples of this kind of resources are the Pathosystems Resource Integration Center (PATRIC) (Gillespie et al. 2011), the Virus Pathogen Resource (VIPR) at Virginia Bioinformatics Institute, and the Comprehensive Microbial Resource (CMR) at the J. Craig Venter Institute (Davidsen et al. 2010).

The information stored in these resources should be analyzed carefully in order to correctly translate it into immobilized probes. The quality of the sequence annotations is a very important parameter in selecting a particular sequence as a template for probe design. In addition to annotation quality, there are several other parameters that should be considered during the selection of biomarker targets and sequences for the design process. The total number of sequences available for a particular target of interest as well as the "completeness" of the available sequences are very important parameters that should be taken into account when selecting a target and designing the probe.

Two major scenarios dealing with the different characteristics of the probe design process can be considered. The first deals with experiments focused on the detection of pathogens of interest, including genotyping, single-base differences, and even resequencing. The second scenario is the better-known approach based on gene expression studies. In both cases, there are differences in the design of the probes depending on the technology used for the manufacturing of the microarrays, since not all of the possibilities are available with every manufacturing platform.

4.2.1 Probe Design in Microarrays for the Detection of Pathogens

Microarray-based experiments can be developed for the detection and characterization of foodborne pathogens. In this scenario, the purpose of the experiments is to identify the presence of one or more pathogens in a particular sample. To accomplish this, it is necessary that probes able to identify only the sequence belonging to the pathogen of interest are immobilized on the surface of the microarray. For the design of the probes, it is possible to divide the experiments into two different categories and then generate different subdivisions:

1. Microarray surface or solid-phase identification reactions. All the experiments that directly use surface probes to report the presence or absence of a particular biomarker coming from a pathogen or pathogens of interest fall under this definition. In this case, it is necessary to develop and design specific probes for each of the biomarkers of interest. This category can be subdivided, as well, depending on the kind of biomarker that is to be detected:

   (a) Gene set analysis. The purpose of this type of analysis is to detect one or more biomarker genes belonging to the organisms of interest and, based on these tests, report either the presence or the absence of the pathogenic organisms. In many cases, a set of probes based on different regions or sequences are used to ensure the detection or identification of a particular organism. It is also common in this approach to focus on meaningful phylogenetic genes such as 16S RNA or gyrases in the case of bacterial analysis (Jarvinen et al. 2009; Kakinuma et al. 2003; Kostic et al. 2007, 2010).
   (b) Single-base analysis. The probes designed for this approach are based on the detection of the difference in a single base between two or more sequences belonging to different strains or subspecies of the same organism. This approach is commonly used in typing assays (Dotsch et al. 2009).

2. Other identification reactions. This type of microarrays is used solely for identifying a positive reaction performed somewhere else, like a PCR or LCR (ligase chain reaction). These approaches most often rely on the use of reporter or universal probes that are pathogen-independent (Cooper and Goering 2003; Ember et al. 2011). The approach based on using universal probes will be described later in a different section in this chapter.

Another very important aspect that must be considered during the design step is the kind of reaction performed on the surface of the microarray, whether the direct detection of the pathogens or the detection of a previous reaction. This parameter can be divided into two major approaches that cover most of the published methods:

• Hybridization-based experiments. The probe design must be focused on the analysis of the hybridization parameters and characteristics, so that all the designed probes work efficiently under the same hybridization conditions.
• Enzymatic-based experiments. Under these conditions, the probe design is driven by the enzymatic reaction that is going to take place on the surface of the microarray, and thus the primary focus is on the requirements for the enzymatic reaction.

Microarray manufacturing technology also has a very important impact on probe characteristics. In some cases, different manufacturing technologies place limits on the use and application of certain experimental approaches and therefore their associated probe designs. Most of the major commercial arrays are manufactured with probes designed for hybridization-based strategies, although there are some technologies where it is possible to order some modifications and alterations that may open the door for different applications.

4.2.1.1 Hybridization Probe Design

Hybridization-based applications are the most commonly used microarray applications, mostly due to gene expression experiments but also due to experiments developed for the identification of pathogens. At the development stage, there were two different approaches for hybridization-based experiments depending on the microarray manufacturing technology. Affymetrix's technology allowed that company to develop high-density microarrays using in situ-synthesized short oligonucleotides, while other companies and most of the "in-house"-built microarrays use PCR products spotted and immobilized on the microarray surfaces. There were different reasons for the use of PCR products, such as cost of oligonucleotide synthesis and sequence availability. As long as the price of oligonucleotide synthesis and the number of available genomes and sequences continues to increase, there has been a shift in microarray manufacturing toward the more widespread use of oligonucleotide arrays, and nowadays, immobilized oligonucleotides are the most common applications. The length of the immobilized probes have varied with time as well and range from 10-mers up to 90-mers depending on several characteristics, such as the application or the technology utilized for oligonucleotide synthesis. Nowadays, probes with sizes ranging from 10-mers to 30-mers are considered short, while 40-mers to 90-mers are considered long.

For microarrays designed for the detection of pathogenic organisms, sensitivity is obviously a key aspect to keep in mind during the design of the whole experiment. In the case of microarray hybridization experiments, in many cases, the probes must work in complex mixtures, where the target abundance may be limited and in a much lower concentration relative to other molecules present in the sample. The probe sensitivity is related to its length and the effective accessibility of the probe for the target. Increasing the length of the oligonucleotide probe increases its sensitivity, because during the hybridization, a longer probe is able to form a more stable hybrid with its target, which generally has a higher binding energy. For example, Chou et al. found that 60-mer probes show an eightfold increased sensitivity with respect to 25-mer probes (Chou et al. 2004). However, there are other aspects regarding probe sensitivity that should be considered. For example, the probe structure and its ability to self-assemble into stable secondary structures such as stem-loops, hairpins, or probe–probe duplexes affect the sensitivity and may reduce it significantly. The most recent probe design software makes use of thermodynamic calculations in order to establish the binding energies for probe–target hybrids, any internal secondary structures, and possible probe dimers. Despite the efforts done in the development of software and models to understand hybridization thermodynamics, most depend on models developed for reactions in solution, and the conditions on the microarray surface might be very different due to the probe density and proximity to the surface. For this reason, these calculations should be treated only as an approach even when their use provides an improvement in the probe design process compared to older, purely sequence-based software that does not take these parameters into consideration. Similarly, much care should be put into establishing different cutoff values for these parameters for probe selection or prediction of its quality.

There is a final aspect related to probe sensitivity that is not directly linked to probe design itself but is related to the probes, their behavior, and their sensitivity. Probes are attached to the surface of the microarray in different ways depending on the chemistry used for their immobilization and the manufacturing processes. In most applications nowadays, the probes are covalently bound by one of their ends, and depending on the way that linking process is done, the surface might be too close to the active site of the probe, which thus causes steric hindrance on the accessibility of the target to the probe. This steric hindrance is due to the lack of space for the target to fold its flanking regions where it is recognized by the probe because of the proximity to the surface. For this reason, it is common to add a spacer region consisting of a stretch of nucleotides to every probe. In many cases, this stretch of nucleotides consists of a 15-mer poly(T) or poly(A) sequence, although other sequences and lengths have been proposed and used.

Together with the sensitivity, specificity is the other major probe characteristic that is of great importance for probes designed to be part of a microarray. Specificity affects the final results at least as much as, if not more than, sensitivity in microarray experiments. Specificity is the ability to recognize solely the correct target sequences that are entirely complementary to the probe. Therefore, it must be taken into consideration along with the balance between perfect matches and mismatches consisting of probes and any nontarget sequences.

As with sensitivity, there is a relationship between probe length and specificity. However, contrary to what happens with sensitivity, specificity decreases as the oligonucleotide probe length increases. The reason for this characteristic of probes is related to the stability of the targetprobe duplexes. When long probes are used, the stability of these complexes is high enough to tolerate mismatches within the sequences, resulting in potential cross-reactivity. The effect of mismatches varies depending on their abundance, their location, and the total length of the probe. The destabilizing effect of a single mismatch is greater when it is located in the central region of the target–probe complex, and its effect is reduced when the mismatch is toward the end of the complex. It is equivalent to a reduction in probe length when it is located immediately at the 3' or 5' end. Similarly, the effect of a set of mismatches is maximized when these mismatches are evenly distributed along the target–probe complex rather than when they are all clustered together in a single region of the probe. The fact that mismatches have a huge effect on the stability of the hybrids and, thus, on the probe specificity points to the possibility of developing probes to detect single-nucleotide variants, for which short nucleotide probes are designed.

The specificity of the probes is analyzed a priori with the use of bioinformatics by performing in silico searches of sequences complementary to the probe with sequence analysis tools such as BLAST or other suffix array methods. This step is contained in many software applications, which also utilize different parameters to offer a measure of cross reactivity. As mentioned previously, it is important to consider the quality of database sequence annotations when utilizing these applications. Comparisons of specificity are done in many cases with the use of databases generated from either the set of targets to avoid cross-reactivity among expected targets represented in the array or with the use of databases or sequence datasets provided by the users. For the correct usage of these algorithms, it is important to know the complexity of samples used for the microarray in order to select the right dataset of sequences to generate the databases that will be used in the comparisons and the prediction of possible cross-reactivities. A particularly extreme case is the design of zip code or tag probes (Gerry et al. 1999; Favis et al. 2005; Girigoswami et al. 2008). Zip code probes were originally defined as artificial sequences that can be used as probes. Their defining characteristic is that each zip code is unique. In order for a zip code sequence to be used as a probe, it must fulfill all the requirements described for any other probe but also must not be found in nature. It is this additional constraint that makes the zip code sequences unique. More recently, the concept of zip code probes has been extended to use as source sequences from organisms that are distant from those that are to be analyzed in the experiments. For example, in the analysis of human samples, zip code probes have been generated using the genome of the bacteriophage lambda.

Microarray experiments usually require a sample amplification step, which needs to be considered during probe design for gene expression or gene set analysis experiments. Usually, most of the software used in microarray probe design allows the user to perform three major kinds of designs depending on the region where the probes are chosen: 3'-end biased; 5'-end biased; and random-primed. The 3'-end and 5'-end options are based on the region of the template target sequence used preferably for the design of the probes. These are the most commonly available options, but depending on the software, there might be other additional options or the 5'-end biased might not be available. The reason why this template region option is available in the design of the probes is related to different aspects, such as accessibility of the target regions based on the amplification procedure or the abundance of the target region due to the processivity of the enzymes and the possible polarity changes in the amplification procedure. Traditionally, in many eukaryotic gene expression assays, amplification based on oligo(dT) primers result in 3'-end biased designs. It should be remarked that nowadays there are some prokaryotic kits that are based on this same approach by means of an initial polyadenylation step of the prokaryotic RNAs.

Due to the fact that in the final microarray a great number of probes are going to be working under the same conditions, it is necessary to design all the probes under the same set of rules. The rules are chosen in such a way that will enable all the probes to work under the same experimental conditions and are generally based on the selection of some common parameters. The most commonly used parameters to ensure experimental uniformity in the probes are the probe length, melting temperature (T_m), GC content, binding free energies, and degree of cross-reactivity.

Despite the particular differences in the parameters used for probe selection among different probe design software, the main characteristics used in the design

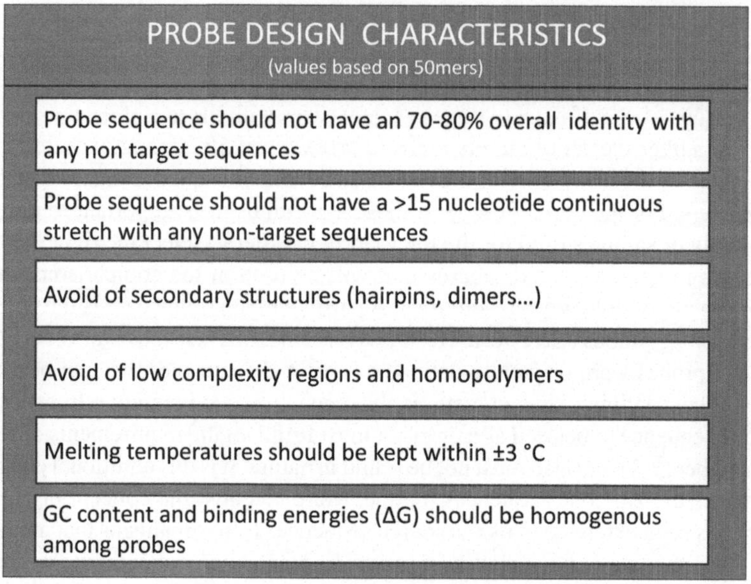

PROBE DESIGN CHARACTERISTICS
(values based on 50mers)

Probe sequence should not have an 70-80% overall identity with any non target sequences

Probe sequence should not have a >15 nucleotide continuous stretch with any non-target sequences

Avoid of secondary structures (hairpins, dimers...)

Avoid of low complexity regions and homopolymers

Melting temperatures should be kept within ±3 °C

GC content and binding energies (ΔG) should be homogenous among probes

Fig. 4.1 Summary of the different parameters that should be considered during microarray probe design

of gene set probes that ensure high specificity are based on recommendations from Kane et al. (2000), which are summarized in Fig. 4.1. An important part of the success or failure of the probes is based on the selection of the right template for probe design.

The design of hybridization probes for SNP analysis, also called *allele-specific oligos* (ASO), is even a bigger challenge. The sensitivity constraints remain the same, but there is an additional reinforcement of specificity constraints due to the fact that the purpose of these probes is to discriminate between single-base variations in the sequences. As stated before, the specificity is affected by the length of the probe, with longer probes being more tolerant to mismatches. Therefore, in SNP analysis, only short probes (11–30 nt) are used. The most well-known approach for SNP analysis has been developed by Affymetrix using sets of 25-mer probes with the SNP of interest located in the middle of the probe. These arrays discriminate between perfect match (PM) and mismatch (MM) hybridizations at each position of interest. As previously discussed, this kind of approach relies on the fact that the disruption caused by the existence of a mismatch located in the middle of a probe provokes a loss in efficiency of the hybridization to a degree high enough to be detected (Fig. 4.2). This approach includes some risks associated with the specificity of the reactions. With the Affymetrix platform, it has been reported that in some cases, the signal coming from MM probes is at least as intense as the signal from PM probes. These kinds of artifacts are removed afterward during data analysis, and those probes are discarded. Despite the location of the mismatch, another factor affecting the specificity of this kind of probe is the nucleotide content where the

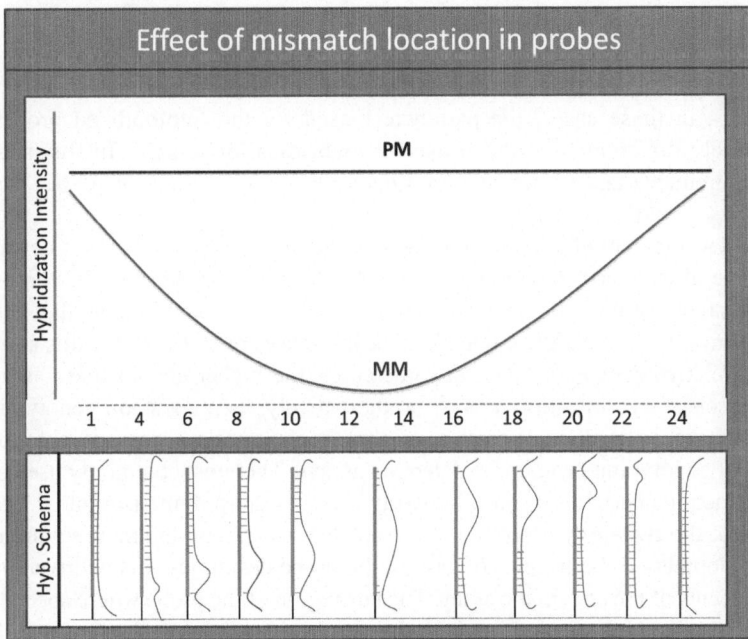

Fig. 4.2 Graph showing the effect on hybridization intensity of the position where the mismatch occurs within the probe. The *lower* part of the graph represents the hybridization and the strength due to hydrogen bonding between the probe and target. The minimal intensity and maximal difference between a perfect match and a mismatch occurs when the mismatch is in the central position of the probe

expected mismatch occurs. Different base mismatches render different hybridization affinities and, thus, different final signal intensities. Similarly, in those cases where the genetic variations being assayed are insertions or deletions, the location of the studied position as well as the adjacent nucleotide environment have an effect, since some structures are more stable than others, resulting in smaller differences in the measured intensities. Unfortunately, many of these parameters have been studied in depth and have been identified as affecting hybridization efficiencies, but they are yet to be implemented into probe design software.

One final approach to hybridization probes is tiling arrays. In this approach, a whole set of probes is designed to span the entire given sequence. This approach commonly uses a complete genome as template. Tiling array probes are designed using the same parameters described previously but with some new parameters included during the design process depending on whether the tiling process is based on distinct probes or overlapping probes. When designing unique probes, it is necessary to define the gap between two consecutive probes. When the design is based on overlapping probes, the adjustable parameter is the tiling phase between probes. The tiling phase is similar to the gap parameter used in the design of unique probes, as it represents the number of bases in the sequence before the probe overlap starts between two adjacent probes in the sequence space.

4.2.1.2 Enzymatic Probe Design

Microarrays can be used as the substrate for the performance of enzymatic reactions; in these cases, the parameters used for the immobilized probes are completely different than those used in hybridization assays. In this section, the requirements and parameters considered during the design of probes for this kind of assay will be described.

The development of enzymatic assays on the microarray surface was developed sometime after hybridization approaches were already established. In contrast to hybridization assays, enzymatic assays are much less common in the field of microarrays. It can be argued that some of the causes responsible for this situation are the relative complexity and, in some cases, the higher cost of these assays in comparison to hybridization assays, or alternatively, their focus on genotyping or SNP analysis. Regardless of the reason, these techniques are very valuable for the design of genotyping or microbe detection assays. The most commonly used enzymatic reactions carried out on microarray surfaces are polymerization or ligation assays; of the two, polymerization, or extension, assays are the more common. In the development of both kinds of assays, the probe design process is driven by the requirements of the enzymatic assay. The key region of the probe where special care should be taken in the design is where the enzymatic reaction will take place. Thus, it is important to realize that a difference between this kind of probe and hybridization probes is that in enzymatic assays, the probes are covalently bound to the surface in a particular orientation (with either the 3′ end or the 5′ end exposed and accessible for the enzymatic reaction). Additionally, as with hybridization probes, a spacer sequence is required to prevent steric hindrances. In contrast to hybridization probes, the binding of the probes to the surface and their orientation are dictated by the manufacturing technology rather than for biological reasons.

In the design of probes for polymerization assays, which are most commonly called *extension assays*, there are two different possibilities based on the kind of extension reaction. In most cases, these experiments rely on the capability of a polymerase to distinguish between a perfect match and a mismatch when elongating an immobilized probe. Under such circumstances, the designed probe must be allele-specific, with the 3′ end of the desired allele to be extended in case of a perfect match. These kinds of probes are therefore closely related to PCR primers in their design but have the additional constraint of limiting the primer to only have the desired 3′ end.

Another method to design primer extension probes is found in the design of APEX (arrayed primer extension) probes. The major characteristic of these assays is that they combine Sanger sequencing and the use of ddNTPs with microarrays. In APEX experiments, the probes are designed to finish one base before the position under analysis. The allele determination is arrived at with the addition of the corresponding fluorescent ddNTP, followed by its detection. Once again, the design of these probes is very similar to the design of allele-specific probes for primer extension.

For both methods of primer extension, the most sensitive region of the probes is the 3′ end, where the enzymatic reaction is to take place and where the assay specificity is determined. In some cases, when an additional SNP or mutation is very

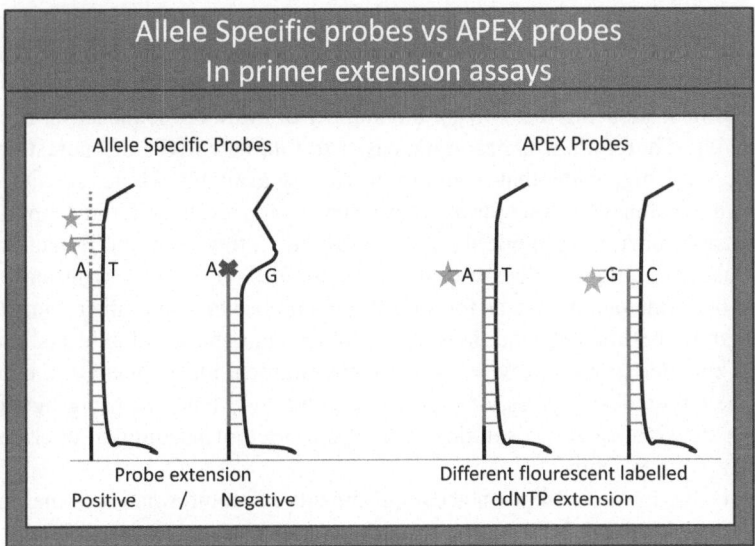

Fig. 4.3 Graph showing the schema of primer extension reactions on microarrays

close to the analyzed position such that they appear together on the same probe, it is possible to design the probe to allow the presence of that mismatch as long as it is not too close to the 3′ end of the probe (Fig. 4.3).

The other seldom used option available for microarray enzymatic assays is the use of ligation assays with one of the ligation oligonucleotides immobilized on the microarray surface. These experiments allow for two different possibilities for their design and in the kinds of probes immobilized on the surface, which depends on whether the immobilized oligonucleotide end has the 3′-OH or the 5′-PO$_4$ exposed and accessible for the ligation reaction. The parameters used in the design of these probes are the same as those used for the immobilization of ligation assays.

4.2.2 Other Aspects Related to Probe Design

Probe design is a key determinant in whether a microarray experiment succeeds or fails. It is influenced by several factors that have already been mentioned, such as manufacturing technology, experiment objectives, and experimental approach. Plus, there are additional points not included in these three aspects that are of interest for the design of the microarrays, such as, for example, how a microarray can be expanded to incorporate new probes, what approach for changing organisms or techniques for the analysis is the best, and what online resources dealing with probes, if any, are of interest. In this section, these questions will be partially discussed along with some examples of probe design for different approaches that use different software.

Sometimes in the design of a microarray, an initial set of organisms is chosen for various reasons, such as the relevance of the organism to the experimental objective, or the amount of available information on the organism. With the advances in sequencing technology, the number of available sequences is increasing quickly. Therefore, it is becoming easier to find new SNPs of interest or new sequences for organisms that were not previously available. Under such circumstances, extension of the microarrays may be necessary, and the different approaches to extension reactions provide different solutions to this problem. Expanding an array based on unique probes would require the design of the new probes using exactly the same parameters as for the previous probes to ensure their compatibility with the experimental conditions. On the other hand, the use of universal probes, such as zip code probes, provides an open environment independent of the organisms, the targeted sequences, or even the methodology (such as using ligation or primer extension assays in solution), but at the price of needing an intermediate step that requires setup.

One of the most well-known accomplishments of bioinformatics in the field of biology and in research laboratories has been the development of different databases for data and information storage and making them available online for the scientific community. Probe design has also benefited from these tools, and it is possible to find several online resources where annotated probes are available. These resources can be catalogued into two different categories:

- General microarray resources. Under this category, we find the major microarray databases where whole experiments are published. In these databases, the experiments are described in detail, including the probes used and their major characteristics.
- Specific probe resources. These resources only store information about probes and their annotations. In most cases, these databases are focused on certain targets such as 16S RNA or only specific organisms. In general, these databases are smaller than the large general microarray resources, but they provide a more comprehensive approach for designing probes for genotyping. An exception is the NCBI's "Probe database." This is a large database where different kinds of probes (not only ones designed for microarrays) are stored and publicly available. Microarray probes are mostly obtained from the NCBI resource for microarrays.

Figure 4.4 contains a list of resources where information about probes for different organisms and different approaches can be found.

As has already been discussed, there is great variety of approaches for the design of probes for microarrays. Similarly, many different types of software have been developed for this purpose. Since gene expression represents the vast majority of microarray applications, most of the available software has been focused on providing different approaches to this type of research. Despite having different interfaces, the major difference between these tools is in the parameters used to define probe specificity. Table 4.1 provides a list of some of the currently available software devoted to probe design with links to the sites where they can be downloaded or

Probe information related resources		
Resource Name	**Purpose**	**Availability**
Gene Expression Omnibus (GEO)	General repository about microarray experiments at the NCBI including probe annotation	http://www.ncbi.nlm.nih.gov/GEO
ArrayExpress (AE)	General repository about microarray experiments at the EBI including probe annotation	http://www.ebi.ac.uk/arrayexpress
CIBEX	General repository about microarray experiments at the DDBJ including probe annotation	http://cibex.nig.ac.jp/index.jsp
Stanford Microarray Database (SMD)	General repository about microarray experiments at Stanford University including probe annotation	http://smd.stanford.edu/
NCBI-Probe	Probe (not only for microarray technologies) repository at the NCBI. Contains few microarray probes for viruses or bacteria	http://www.ncbi.nlm.nih.gov/probe/
ProbeBase	Focused on 16S RNA probes	http://www.microbial-ecology.net/probebase/
BμG@S	Contains information about microarrays for certain microrganisms and their annotated probes.	http://bugs.sgul.ac.uk/
OligoArrayDB	Contains probes designed using OligoArray software for more than 900 organisms or strains (mostly bacteria)	http://berry.engin.umich.edu/oligoarraydb/
Viral Probe Database	Contains probes designed for viruses	http://genestamp.sinica.edu.tw/virus/

Fig. 4.4 Table containing information resources for microarray probes

Table 4.1 List of probe design software options and availability

Software name	Operating system[a]	Purpose	Availability
AlleleID	W, M, L	Hybridization[b] Enzymatic probes[b] Allele-specific probes[c]	Free demo/commercial http://www.premierbiosoft.com
ArrayDesigner	W, M, L	Hybridization[b] Enzymatic probes[b] Allele-specific probes[c]	Free demo/commercial http://www.premierbiosoft.com
ArrayOligoSelector	L	Hybridization probes[b]	Free for academia/downloadable http://arrayoligosel.source-forge.net/
CommOligo	W	Hybridization probes[b]	Free for academia/downloadable http://ieg.ou.edu/software.htm
eArray	O	Hybridization probes[b]	Online/Agilent's microarrays https://earray.chem.agilent.com/earray/
GoArrays	L	Hybridization probes[b]	Free for academia/downloadable http://g2im.u-clermont1.fr/serimour/goarrays.html

(continued)

Table 4.1   (continued)

Software name	Operating system[a]	Purpose	Availability
HiSpOD	O	Hybridization probes[d]	Free for academia/website http://fc.isima.fr/~g2im/hispod/page_about.php
OligoArray	W, M, L	Hybridization probes[b]	Free for academia/downloadable http://berry.engin.umich.edu/oligoarray2_1/
OligoFaktory	M	Hybridization probes[b]	Free for academia/downloadable www.bioinformatics.org/oligofaktory
OligoPicker	L	Hybridization probes[b]	Free for academia/downloadable http://pga.mgh.harvard.edu/oligopicker/
OligoWiz	W, M, L	Hybridization probes[b]	Free for academia/downloadable http://www.cbs.dtu.dk/services/OligoWiz/
Visual OMP	W	Hybridization probes[b] Allele-specific probes[c]	Commercial http://dnasoftware.com/
Osprey	O	Hybridization probes[b]	Free for academia/website http://osprey.ucalgary.ca/
PhylArray	O	Hybridization probes[b]	Free for academia/website http://g2im.u-clermont1.fr/serimour/phylarray/about.php
PICKY	W, M, L	Hybridization probes[b]	Free for academia/downloadable http://www.complex.iastate.edu/download/Picky/index.html
ProbeTool	W, L	Hybridization probes[b]	Free for academia/downloadable http://jakob.genetik.uni-koeln.de/bioinformatik/software/probedesign/index.html
Probe Designer	O	Allele-specific probes[c] APEX probes	Free for academia/website http://www.biodata.ee/cgi-bin/probe_designer/index.cgin
ProDesign	O, L	Hybridization probes[b]	Free for academia/website/downloadable http://www.uhnresearch.ca/labs/tillier/ProDesign/ProDesign.html
ROSO	O, W	Hybridization probes[b]	Free for academia/downloadable http://pbil.univ-lyon1.fr/roso/Home.php
UPS2.0	O	Hybridization probes[b]	Free for academia/website http://array.iis.sinica.edu.tw/ups/

[a]W (Windows), M (Mac), L (Linux), O (online)
[b]Gene expression/gene set analysis
[c]SNP analysis
[d]Gene set analysis

used online. Only a small number of the software applications are able to design probes for genotyping. Not included in this list are primer design tools that can be used for the design of probes as well. Most of the software suggested in this list is freely available for academic institutions and universities.

4.3 Microarray Data Analysis

Microarray data analysis represents a major challenge in microarray experiments, as it is a big part of the experiments. There are different methods of carrying out the analysis depending on the approach used in the experiments. The focus and design of analysis are based on the experimental objectives. In this section, we will provide an overview of the available possibilities and analysis strategies for genotyping and gene expression–based approaches. Microarray data analysis can be divided as follows:

1. Image analysis. Microarray analysis really starts with the analysis and quantification of the image acquired after the reaction on the microarray surface. In this step, images are transformed into numerical values to be used afterward in the numerical analysis.
2. Numerical analysis and visualization. Commonly considered as the only step of data analysis, the procedures applied will be focused on the interpretation of the numerical data generated in the image analysis. Depending on the experimental approach followed, different analyses will differ in the tools used and the objectives; for instance, different analytical methods are used for gene expression analysis as for genotyping and, therefore, both possibilities will be discussed.

Microarray data analysis is a complex process comprised of a set of several sequential steps, each of which requires different algorithms and has different effects on the final experimental results. Data analysis has been a hot topic in the fields of microarrays and bioinformatics, generating a huge amount of software and bibliography, primarily focused on gene expression.

4.3.1 Microarray Image Analysis

Image analysis is the first step in data analysis since through the use of scanners or other imaging technologies it translates the images generated during the experiments into numerical values corresponding to fluorescence intensities to be used and stored afterward. In many cases, the step of image analysis has not received the required attention by researchers who rely blindly on the software regardless of the experiment performed or the platform used. It should be kept in mind that this step is where data are originated and any bias introduced will continue in any further analysis; therefore, the quality of this stage will determine the quality of the subsequent steps and the final conclusions of the experiment.

Image processing and analysis issues can be separated into several sections: gridding, segmentation, quantification, and, finally, quality assessment. There are a great variety of image analysis tools providing an integrated interface for the user and requiring that the user defines a different number of parameters related to these steps depending on the software and platform used.

Microarrays contain a structured and organized grid of biological probes' immobilization on a solid surface. During the gridding step, this pattern of probes with the information regarding the position of each surface is used to identify spots on the image. During the gridding process, an ideal microarray model is aligned with the scanned microarray image. Depending on the software chosen, the user might be required to set some basic parameters, such as

- The number of files and columns of the array and the distance between spots.
- The number of elements in the array, referred to as metarows and metacolumns, or subarrays in the case of more than one array.
- The spot size diameter in pixels. It is necessary to be aware of the resolution (pixel size in micrometers) used during the scanning process.
- Probe annotation data, such as probe name, sequence, function, etc.

Depending on the software, it might also be necessary to adjust the position of the defined grid on the image to allow the system to identify the spots. This is done by adjusting each position according to the real position on the image by correcting some imperfections and deviations with respect to the ideal spot distribution due to manufacturing issues. In some other cases, the system is able to set the grid location automatically on the image without any further assistance from the user. In any case, it is recommended that the user checks the adjustment of the grid (Fig. 4.5).

Once the gridding step is done, the next step in image processing is the spot segmentation. Spot segmentation together with the gridding step is the key for successful image analysis. It is in this step where the image pixels corresponding to a spot are classified as belonging to a signal due to a positive reaction of the sample with the probe or, conversely, background noise coming from nonspecific source reactions. The source and amount of background in the experiments may originate from very heterogeneous factors, such as the substrate material, washing steps, or manufacturing processes. The importance of this step has fostered the development of different methods and algorithms for an improvement in the quality of the results. Today different software tools are available for this purpose that employ different algorithms and strategies for the segmentation. As mentioned before, image analysis is a sequential process, and the starting point for the segmentation of the images is the gridding step. Segmentation algorithms are applied on those regions where the spots are located during the gridding, and these algorithms can be classified as either spatially based or intensity-based methods.

Spatially based segmentation methods were the first segmentation algorithms applied to microarray image analysis, and they rely on both the position and quality of the spots. In these methods, correct gridding is essential to ensure optimal performance since they are based on the central position of the theoretical gridding position and to a lesser extent on the underlying image parameters. Among these

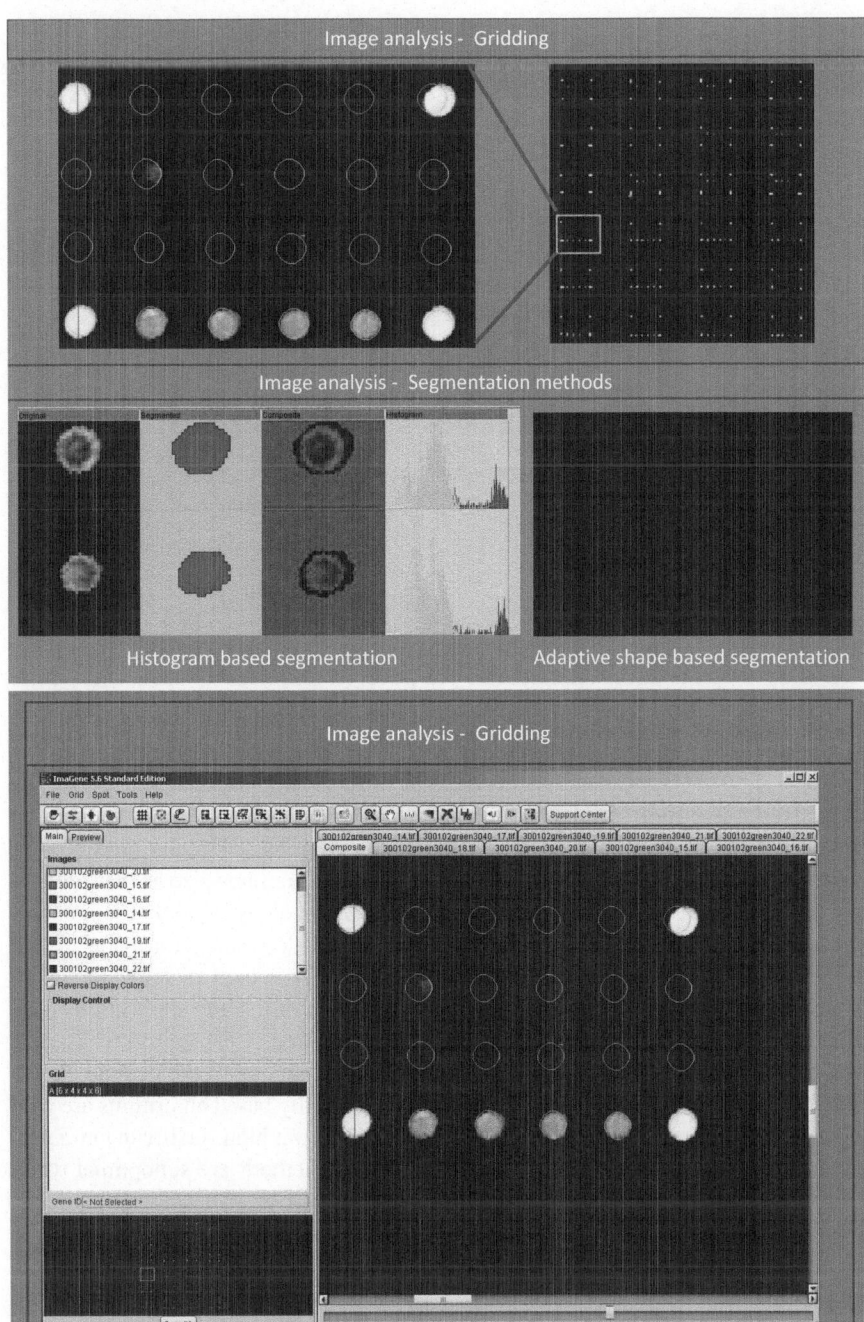

Fig. 4.5 Example of a gridded microarray and image showing the differences between different segmentation algorithms

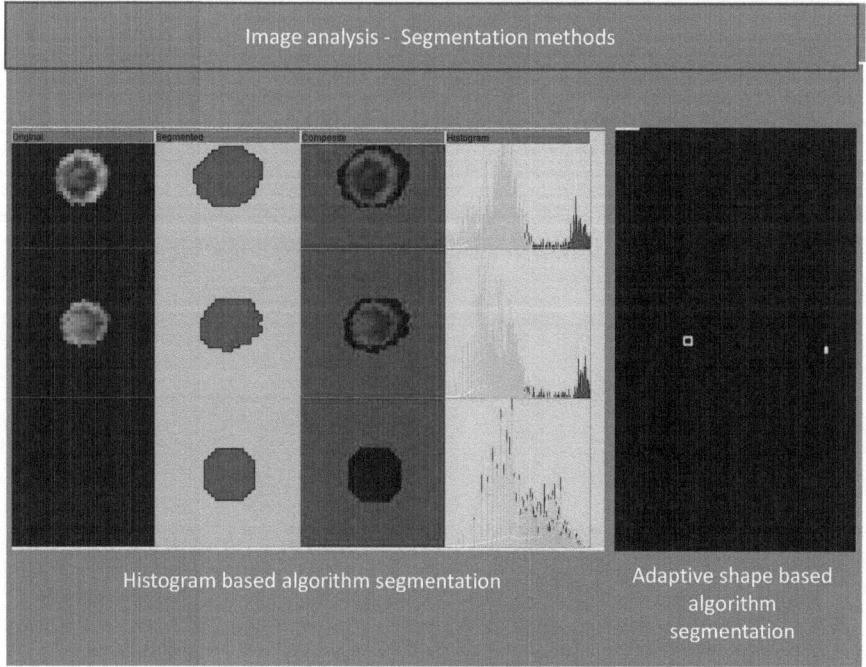

Fig. 4.5 (continued)

methods, the two most common types are based on a fixed circle and on adaptive circles. In both of these methods, the segmented region has a circular morphology. In the fixed-circle method, the user defines a diameter the same size as the spot that will be considered the "signal" section, and the pixels belonging to that region will be considered "signal pixels." A second circle with a wider diameter is made around the first. The region between the inner and outer circles is used to define the background region. This method is still recommended today for some applications. The adaptive circle algorithm was an improvement on the fixed-circle approach based on the possibility of allowing the system to modify the diameter of the defined circles as the spots that are not all homogeneous in size. Spatially based algorithms are well suited for homogeneous spots, such as those coming from high-quality manufacturing processes, and have a good round shape. These methods are suboptimal when the spots are not perfectly round, like those oval in shape or with irregular borders.

Intensity-based segmentation algorithms were developed to overcome some of the inefficiencies and drawbacks of spatially based algorithms. These methods allow the identification and segmentation of spots with irregular borders and likely define the real morphology of the spots and eliminate some artifacts that might appear in the images. The utilization of these methods relies on the assumption that there is a statistically significant difference between signal and background pixels. As with the spatially based methods, there are different approaches and different algorithms for intensity-based methods. In one approach, called a histogram-based approach, a threshold value is defined and the assigning of a spot as either signal or background is based on whether the intensity of the pixel is higher or lower than the threshold value. These algorithms have some advantages and drawbacks related to the fact that they do not use any morphological information from the spots (Fig. 4.5). A second category within the intensity-based algorithms is those based on adaptive shapes or growing regions. In these algorithms, the segmentation is based on the use of seeds that are placed in the regions where the spots are located and then the intensities between those seed pixels and their neighbors are calculated. Depending on some statistical parameters of these differences, the region keeps growing and assigning new pixels to the signal region until no more pixels can be added.

Once segmentation has been carried out, the next step is data quantification. In this step, the fluorescence intensity is transformed into numerical data that will be used as values representing either signal or background, based on the analysis of the fluorescence intensity of the segmented regions. There are different ways to generate this numerical data, such as the mean value of the pixels, the median, the total intensity, or the sum of all intensities. Since there are multiple pixels in each of the segmented regions, it is possible to perform various statistical analyses. These statistical analyses, such as standard deviation or coefficient of variation, can be used afterward to run a quality assessment on the spots. Several factors may affect image quality, image segmentation, and data quality. At this time, it is possible to identify some of these problems and label them with a process called *spot flagging*. Flagging differs depending on the quantification software and is automatically done, in many cases, by the software. The user may be allowed to add some "manual flags" to identify any possible artifacts in the image not identified by the software. Some of the most common problems that are identified and therefore flagged are

- Spot saturation. This is a common flag used when the intensity of the spot is above the scanner's detection limit (generally, values above 65.565). At this point, the flag indicates that the intensity of the spot might be higher that the limit but cannot be quantified, and therefore, the spot should be treated carefully and generally filtered out for further analyses. To avoid these issues, microarrays are scanned under different conditions to avoid the presence of saturated spots. However, despite such efforts, sometimes it is impossible to avoid the presence of some saturated spots.
- Spot morphology and size. With this flag, the software identifies spots that differ from the ideal round morphology. In other cases, the size of the segmented area defined as signal is different from the expected size.
- Controls. This flag can be defined by either the manufacturer or the user to easily identify the controls present in the microarray experiments.
- Positive, blank, or empty spots. The goal of these flags is to make an automatic annotation depending on whether the spots have been detected or not. There are different approaches for determining whether a spot is present or not (i.e., a positive reaction). Some of the most commonly used algorithms for this purpose are either based on statistics, like a t-test between the intensities of signal and background pixels using different p-value limits (generally, 0.01 or 0.001), or by using the ratios between signal and background intensities (generally applying a value >3).

Table 4.2 List of some image analysis tools

Resource name	Characteristics	Availability
Bioconductor	Bioinformatics repository with scripts developed for R language. Some tools designed for image analysis. Multiplatform	http://www.bioconductor.org
Scanalyze	One of the first software developed for microarray image analysis. It is focused on the analysis of dual hybridization experiments. Windows	http://rana.lbl.gov/ EisenSoftware.htm
TIGR spotfinder	Designed to quantify dual hybridization experiments. Multiplatform	http://www.tm4.org/spotfinde. html
Spatter	Designed to quantify dual hybridization experiments. Linux	http://www.nongnu.org/spatter/
AGScan	Designed for image quantification. Multiplatform	http://mulcyber.toulouse.inra.fr/ projects/agscan
UCSF spot	Designed for image quantification	http://www.jainlab.org/ downloads.html
MAIA	Designed to quantify dual hybridizations	http://bioinfo.curie.fr/projects/ maia/
GenePix Pro	Commercial/automatic gridding	Molecular Devices http://www. moleculardevices.com
Feature extraction	Commercial/for Agilent's and other slide arrays	Agilent http://www.agilent.com
Phoretix array	Commercial/designed for protein arrays	TotalLab http://www.totallab. com
ImaGene	Commercial/supports several images at once	Biodiscovery http://www. biodiscovery.com

Once the images have been analyzed and numerical values generated, the results are generally saved in tabular format or spreadsheet files with the gridding annotation and the numerical values associated with the quantification. These files are going to be the starting point for any further analysis. Table 4.2 contains a list of some software tools for image analysis.

Although it will be discussed in further sections of this chapter, it can already be pointed out that from image analysis, only the general protocols, software used, and the final quantified files are submitted to public databases as annotations to the experiments. Neither images nor details on the actual algorithms used are uploaded into the databases.

4.3.2 Microarray Numerical Data Analysis

In most cases, microarray data analysis is carried out only once numerical data from the experiments are available. Therefore, in this chapter, data analysis will refer only to the task of numerical data analysis. In contrast to "wet bench" work, where genotyping microarray experiments might be more time-consuming or more difficult than gene expression experiments, when it comes to dealing with numerical data analysis, the opposite is true, as, generally, gene expression analysis is more time-consuming than genotyping analyses. Most data analysis applications have focused on gene expression experiments, where an explosion of data associated with microarrays has occurred. These applications have also been the most widely used. Nevertheless, there are also some issues that must be considered with regard to the use of microarrays in genotyping and other kinds of assays and specifically, their data analysis.

4.3.2.1   Gene Expression Data Analysis

It has been stated several times throughout this chapter that gene expression analyses are the most common in the field of microarrays. These approaches are mostly based on quantitative strategies. In general in these experiments, it is not enough to know whether a gene is expressed but also to what degree it is expressed. Several of the traditional approaches to statistical analysis cannot be used with gene expression assays due to the fact that the number of variables largely exceeds the number of study cases by at least one or two orders of magnitude. Under such circumstances, it is necessary to proceed carefully and provide results with the appropriate corrections.

This section provides an overview of gene expression data analysis, focusing on two major topics related to microarray analysis:

*   Data preprocessing. All the steps necessary to prepare numerical data for statistical analysis are considered data preprocessing. It is an important step to ensure the quality and significance of the results obtained in the next phase of the analysis.
*   Statistical analysis. Based on this analysis, numerical data are used to draw conclusions from the experiments. There are different approaches to statistical analysis depending on the purpose and the biological questions posed.

Other quantitative-based approaches, such as those used in protein microarrays, follow the same analysis pipeline, including some small differences in particular issues associated with their different platforms.

The protocols and methodologies applied in data analysis should be carefully annotated, since they will be required by the microarray experiment annotation standards required for publication of the data and results.

Data Preprocessing

During gene expression data analysis, the first step is data preprocessing. It is comprised of different tasks whose main goal is preparing the data for further statistical analysis by reducing noise and sources of variation. Data preprocessing and filtering are important because the data are prepared for biologically driven analyses, and all undesired experimental artifacts are removed, such as different hybridization effects and noises, differences in sample source and processing, and microarray manufacturing artifacts (Herrero et al. 2003; Rotter et al. 2008). As usual, in all these processes, it is possible to identify different steps that should be focused on with different degrees of complexity, and the effect each has on the final results may vary from insubstantial to large and highly important.

- Spot quality filtering. Here, quality assessment of the image analysis is carried out for each spot on the microarray, removing spots labeled (flagged) as poor quality for any of the available options used during image analysis. This step is obviously affected by the image analysis software and the parameters selected. Including poor quality spots in the analysis may lead to an increase in the noise of the following statistical analyses and, therefore, a lack in confidence of the final results.
- Background correction. In the experiments, there are some values associated with background correction for each of the measurements done on the spots. This background is subtracted in some way from the signal values in order to identify only the portion of the signal coming from the interaction between probe and target. There are different approaches and strategies for this subtraction depending on the microarray platform, the assumptions made with regard to the background effects on the real signal, and also, the different models developed to correct for these effects. The simplest way to correct the signals would be to utilize the assumption that background represents an additive intensity with respect to the signal, and therefore, it would be possible to perform local background subtraction by directly subtracting the intensity of the noise or background segmented area in the images from the intensity of those regions considered to be signal coming from the probe on every spot of the microarray. In other approaches, a more complex model is built based on the assumption that background neither affects all the surface of the array uniformly, nor is its effect additive. In those approaches where the spots lack any associated background values, there are other ways to estimate the background noise based on the surface and other artifacts. For example, in Affymetrix arrays, the value of mismatched probes could be considered an estimate of the background for a perfect match probe. Although background correction is generally accepted and routinely used in many different models, some authors have suggested that this kind of correction is not necessary for some microarray approaches and can be skipped.
- Normalization. This is probably one of the most important steps during data preprocessing. This process attempts to compensate the systematic differences observed among different samples measured during the experiments, such as differences in the samples, their extraction and labeling processes, the different individual microarrays used, or the hybridization or washing procedures, to highlight some of the possible variation sources. During normalization, a set of mathematical formulas is used in such a way that after normalization, the data can be compared without affecting the information content. There are different normalization strategies depending on the experimental design and the microarrays used (Ding and Wilkins 2004; Park et al. 2003; Steinhoff and Vingron 2006).

During microarray transcriptomic (gene expression) profiling, thousands of genes are generally considered. It is assumed that most of them will not be differentially expressed and also that there will be a balance between upregulated and downregulated genes in the array. These assumptions have formed the basis for most normalization techniques (Loess, Lowess, scaling, quantile, etc.), which have been developed to fit this global approach. Therefore, their use in different scenarios may give misleading results and incorrect final conclusions, such as when only a small subset of genes are a priori expected to be differentially expressed. For the selection of an appropriate normalization strategy, it is necessary to take into account some other parameters, such as

- Array structure and manufacturing. Different manufacturing techniques may lead to different normalization strategies. For example, Affymetrix arrays utilize a different normalization strategy than other arrays or printed arrays manufactured with different tips might utilizes, considering each printing tip is different.
- Labeling scheme. Microarray gene expression experiments can be carried out using competitive hybridizations (two different fluorescent-labeled samples per array), also called *dual labeling arrays*, or with just one sample (one fluorescent label) per array. The two approaches use different normalization strategies. In the first case, two different normalizations might be used, an intraslide one and an interslide one, while in the second case, only the interslide normalization is necessary.
- Gene representation on the slide. Depending on whether the array contains one or more probes used to report a single target sequence, the normalization will be different.

Depending on the assumptions made during the design of the experiment and the normalization method, it is possible to group the normalization algorithms into four large categories:

- Global-based methods. These normalization algorithms rely on the concept that the bias introduced during the experiments can be summarized by means of a "global" constant or scaling factor; therefore, the intensities are corrected for global multiplicative effects. There are different ways to calculate the global normalization constant, such as the use of the mean, median, Z-score, or, in the case of Affymetrix's MAS 5.0, a trimmed mean (Autio et al. 2009).
- Error model based methods. In these normalization algorithms, the goal is the correction of any intensity caused by variation by adjusting it to make it constant along the complete range of intensities. Under these methods, it is possible to quantify the differential expression in a way that is independent of the mean intensities. An example of one of these methods is the variance stabilization normalization (VSN), a frequently used normalization algorithm (Bolstad et al. 2003).
- Transformation-based methods. These are the most commonly used methods, especially among dual labeling experimental approaches. In this category of algorithms, the "quantile"-based normalization or the "Loess/Lowess" normalization algorithms are frequently utilized. In the case of the "quantile" normalization, the user accepts the assumption that genes have approximately the same

distribution among all the samples analyzed, and that following the normalization, the statistical parameters such as the mean, median, and percentiles will be identical among all the samples (Do and Choi 2006).

- In some restricted conditions, such as analyzing only a few genes or when most of the analyzed genes are deregulated, the use of different normalization approaches, such as those based on the use of some controls like spike-in controls, is possible. Spike-in controls are external controls that might be used for different purposes like quality assessment and normalization. These controls are mRNA samples that are added to all the analyzed samples at different known concentrations. Therefore, these controls undergo the same processes as the samples, and their final values for each of the assays should be equivalent. This equivalence in values allows their use as normalization controls. The spike-in should be carefully designed in such a way that they are not present nor have homologs in the analyzed samples, so they cannot generate any cross-reactivity with any other gene measured in the sample.

- Data transformation. This step involves a set of different processes that include the logarithmic transformation of the data and the inputting of missing values for a particular probe in a sample. For example, logarithmic transformation is almost always routine in microarray analysis, generally in base 2, while inputting missing values is needed only in the cases where further analysis cannot be done if there are missing values in the analyzed dataset. Another example of data transformations that might be applied at this step is the merging of replicates.

- Data filtering. This is an important step in gene expression data preprocessing. The purpose of this step is to reduce the complexity of the dataset and prepare it for biologically meaningful statistical analysis. In gene expression analysis, the experimenters are interested in those genes that are differentially expressed. Therefore, those genes that do not vary among samples are going to be a confounding factor in further steps of the analysis and should thus be removed. The removal of genes is also called *flat pattern filtering*. There are different ways and algorithms designed for this kind of filtering based, for example, on the number of "peaks" or the standard deviation of the genes among samples.

As we have seen, many options are available for data preprocessing that will affect in different ways the final results of the biologically meaningful analysis. Depending on the analyses to be carried out, the number of samples available, and the sensitivity in differences of gene expression to be detected in the same experiment, it is possible and recommended to change these filtering parameters to give different and complementary perspectives of the available dataset.

Biologically Meaningful Statistical Data Analysis

This is the final step in gene expression experiments. The purpose of all the methods, algorithms, and databases used is to answer biological questions posed during the design of the experiments. At this step, there are different options and approaches for the analysis that can be grouped in different ways. Considering a "point of view" approach, it is possible to group the experiments into two major categories:

- Gene-centered approaches. In these approaches, the analyses are carried out in such a way that the results represent a list of genes differentially expressed to answer the biological question posed. This is the most commonly followed approach during gene expression analysis and the first approach built when microarrays were developed. These genes are selected by different statistical algorithms. This should provide not only the statistical value that was used to select them but also a false discovery rate value. This value will give an idea about the certainty that members of a particular gene list are really differentially expressed. In most situations, the experimenter must make sense of the biological meaning of the genes identified under this approach.
- Biological process–centered approaches. This approach for the analysis of gene expression is more recent than the gene-centered approach. Instead of just using the genes and their values to select a list of differentially expressed genes, the purpose of this approach is the analysis of sets of genes grouped under biological processes that can be defined either by pathways annotated in different databases or by gene ontology terms. In these analyses, the gene sets are analyzed and the differences to be detected are based on those gene sets instead of individual genes. The major advantage of these methods is that they provide a direct view of those biological processes that are affected in the experiments in a way that the experimenter does not need to give an interpretation for the presence or absence of individual genes.

Either approach can be followed if the purpose of the experiment is to identify new characteristics among the samples or if the experiments are focused only on the study of a few subsets of genes or pathways of interest.

The second option to group the gene expression analysis is based on the type of biological question that will be answered during the analysis. There are three major approaches: class comparison, class prediction, and class discovery. Within this classification, it is possible to distinguish two different methodological approaches. This will depend on whether supervised analytical approaches are used in the case of class comparison or prediction and unsupervised analytical approaches for class discovery. In supervised analyses, the user provides information to the systems and algorithms regarding the classes present and which elements in the analyses belong to each of the analyzed classes. On the other hand, unsupervised methods do not use any kind of a priori information associated with the analyzed elements.

In class comparison studies, the user already knows a set of characteristics or classes that are associated with each sample of the analysis. Studies are focused on the identification of those elements (either genes or pathways) that are differentially expressed among the defined classes. These approaches are used not only in gene expression analyses but also in other microarray-based experiments or in tissue microarrays. Under this topic, it is possible to find a large amount of statistical tools that have been used and adapted for these analyses. Some examples of the tests applied in this kind of studies are the t-test, F-test, or ANOVA. It should be kept in mind that due to the nature of microarrays, some of the statistical approaches followed in other assays may fail due to false positives. Considering a p-value of $p = 0.05$ commonly used in other kinds of approaches would easily provide an assay with 10,000 probes; 500 probes are differentially expressed just by chance. Different methods have been used to provide corrected p-values adjusted for these conditions that provide enhanced stringency to the analysis as well as a tool for multiple comparison corrections.

Nowadays, the results generated with these approaches are given using a false discovery rate (FDR) as a measure of their reliability. The FDR provides an insight about the number or probability of detecting some elements as false positives within the resulting list of differentially expressed elements. There are different ways to keep the proportion of false positives retrieved during these analyses under control. Most of them are centered on either control and limit the number of false positives that might appear in a results list or the control of the proportion of false positives within the results list. One of the most commonly used software applications for class comparison and FDR control is the statistical analysis of microarrays (SAM). Currently, most of the analysis software includes one or more tools, in some cases based on SAM, to provide results with these kinds of corrections.

There is a huge amount of software available for microarray data analysis. Most of this software includes some preprocessing elements as well as different implementations of previously described approaches for the statistical analysis of biologically meaningful questions. Although controlling the number of false positives within our results lists is a very important aspect, it should be kept in mind that the efforts done in control of these false positives may affect the false negatives that are not being detected in the analyses as well. This increment in the rate of false negatives associated with the reduction of false positives might become an issue, especially in cases where the number of analyzed samples is small. For those reasons, it is necessary to compromise between both situations.

Class comparison approaches, combined with functional analyses or biological process–centered approaches, are commonly called gene set enrichment analysis (GSEA). Functional analyses try to come up with underlying biological processes associated with the comparison. The goal of these approaches is to identify whether there is an "enrichment" or statistically significant increase or decrease in the abundance of probes associated with certain biological processes. These kinds of approaches were initially focused on the analyses based on the annotation of probes with gene ontology elements. They associate the differentially expressed elements and the probe annotations within the *Kyoto Encyclopedia of Genes and Genomes* (KEGG) (Kanehisa and Goto 2000) or BioCarta. With the development of new available information sources, these comparisons have been extended to other areas such as interactome analysis, analysis of the enrichment in transcription factors, protein domains, or other ontologies or network data sources. To enhance the results when running this kind of analysis, it is recommended to relax the stringency of the filtering parameters for the probes, limiting them to remove missing values or poor quality spots. In cases when these approaches are applied to datasets with a reduced number of probes/genes, the results are not usually satisfactory, since the reduced number of genes is not able to capture the underlying biology of the analyzed processes.

Class prediction is the second of the procedures where supervised methods are applied. This means that the user is going to provide information about the classes or categories that are known among the samples to the algorithms, and the system will use that information during the analysis. In this case, the aim of these analyses is to identify a set of probes that, combined with some rules, can act as a "predictor" or "classifier" that will not only discern between the classes but will also assign an unknown test sample to any of the previously known classes. This is an important approach in clinics and has been developed for other fields, such as prognosis, diagnosis, or prediction of treatment outcomes for cancer.

In class prediction experiments, it is important to split the available dataset with well-characterized samples into two different datasets. One of the datasets is provided to the algorithms that are applied to build "the rules" used to classify the set of samples, which is called the *training set*. This training set is made up of samples known to belong to those classes or categories of interest to be used by the system as the basis to build up the predictor. The quality of the training set is crucial to have a high-quality process of identifying the predictors. These training sets should have a balanced number of samples belonging to each of the classes to be predicted and should represent as much of the intraclass variability as possible. Examples of the most commonly used algorithms are the artificial neural networks (ANN), discriminant analysis (DA), support vector machines (SVM), the K-nearest neighbor (KNN), or Bayesian methods.

The second dataset, called the *validation dataset*, is not used in the process of building up the classifier, but rather in the process of validation. It is important for the correct procedure of this analysis that the validation dataset is not used in any way during the training process. The validation process will qualify the performance of the generated classifiers by using the validation dataset.

In many cases, class prediction experiments have pitfalls due to the small number of samples available for building them. For this reason, it has been necessary to find solutions to overcome this small sample number such as the application of resampling techniques and the development of cross-validation methods such as the leave-one-out cross-validation (LOOCV).

Class discovery is the third of the analytical methods proposed to give biological meaning to the microarray data analysis. This approach is probably the first approach that was used to analyze gene expression data in microarray experiments and also the most common approach used. In many cases, microarray analyses have been biased toward the use of these methods even when their use was not recommended or did not provide any useful information for the posed question. Examples include cases where it is wrongly used in experiments where the posed question should be addressed using either class comparison or class discovery experiments. The name of this approach for the data analysis comes from the fact that the purpose of these analyses is to identify new classes or hidden characteristics in the analyzed data. For this same reason, the methods and algorithms utilized in these analyses are unsupervised, where no a priori information is used nor needed. Many of these algorithms try to "group" the samples or genes by means of some similarity measured somehow. Some examples of commonly used algorithms are the principal component analysis (PCA) and different clustering techniques. In the PCA analysis, genes are combined into linear combinations that account for dataset variability and reduce the dimensionality of the dataset species. In many cases, PCA is combined with graphical approaches where the three principal components are represented in 3D plots. In those 3D plots, the experimenters can check whether the samples are grouped in any particular way or what their distribution is in this reduced space. The clustering approaches are another way of grouping both samples and genes, by far the most commonly used data analysis technique in microarray analysis. In clustering techniques, the samples or genes are grouped together by means of different algorithms relying on different approaches such as the self-organizing map (SOM), K-means clustering (KM), or the hierarchical clustering (HC). The grouping is done depending on the way the user and the algorithm measure the distance among the elements to be clustered. Among the previously cited clustering methods, the most common one is hierarchical clustering. In hierarchical clustering, the results of the algorithms come out in the shape of hierarchical trees, where the elements are grouped by pairs and then linked to another element. These algorithms can be either agglomerative (joining the samples depending on their similarities and following a bottom-up approach) or divisive (dividing the group by detecting the most dissimilar elements following a top-down approach). In these algorithms it is important to distinguish between two terms that are related to the way the distances are measured: the distance metrics and the linkage metrics.

The *distance metrics* is the way the distance between two elements is calculated. These distances can be calculated using methods such as different correlation measures (parametric or nonparametric) or space distances. (In this case, each probe is considered to be a vector with as many components as samples and likewise with samples, each sample is a vector with as many components as probes selected for the analysis).

On the other hand, *linkage metrics* works such that once two elements are considered to be the closest to each other, they are linked together. As with distance metrics, different approaches to linkage metrics can be used. The final aspect of trees generated by clustering approaches may reflect different views of the same dataset depending on the selection of the parameters used and in some cases may look very different from one another when the algorithm has been run using different metrics. The use of clustering algorithms requires, in many cases, the use of some confirmatory techniques that explain the robustness of the clustering process. Thus, for some tools, it is possible to run some complementary tests, like bootstrap, to analyze how repetitive and therefore how reliable the clusters generated during the analyses are.

Clustering analyses are also used as data quality exploratory analyses. It is common to use these analyses before running supervised methods, such as class comparisons, to explore whether all samples belonging to the same category are grouped together, as would be expected, or if it is possible to identify a sample that behaves abnormally for any reason and therefore would be recommended to be removed from further analyses.

A characteristic in all the unsupervised techniques is that the user is responsible for deciding whether a group really does exist or what its importance for the studied process is.

In the last decade, a huge number of tools have been developed for gene expression data analysis. Among these tools, there are some that can be downloaded and used as desktop tools and others that can be used online. Tables 4.3 and 4.4 provide a list of tools available for microarray analysis. It is interesting to highlight the development of the Bioconductor repository among these tools. Bioconductor is an open-source software project for bioinformatics. In this project, it is possible to find a vast number of tools for statistical analysis and graphical representation of genomic data generated from a variety of platforms such as microarrays or next-generation sequencing. Among the set of microarray data analysis tools, there are a plethora of different scripts and applications not only for microarray gene expression data analysis but also for other applications such as comparative genome hybridization or genotyping. Some of these tools have been designed for specific microarray manufacturers and platforms. It is possible to find tools specializing in data analysis, including preprocessing steps for Affymetrix microarrays, Agilent microarrays, or Illumina arrays.

4.3.2.2 Genotyping Data Analysis

Genotyping and resequencing experiments are designed to answer different questions than those posed with gene expression experiments. While in the gene expression experiments the major aim is focused on the quantitative aspects of the generated data, in many other experiments the goal is to identify those probes that are associated with the genotype present in the sample. Ideally, the genotyping approach should only be considered a qualitative analysis focused on the detection of those unique spots with a positive reaction and, therefore, the only spots that will have a detectable signal on the microarray. Unfortunately, the real experiments in many cases are far from that ideal situation due to a multitude of reasons, such as the unspecific reactions that may happen on the microarray. When the experiment is focused on the detection of single nucleotides, the process of assigning the right signal to the analyzed position is termed *base calling*. The complexity of the base-calling algorithms and processes is related to the approach selected to run the experiment.

In the simplest experimental approach, where there is a unique probe for the detection of a target biomarker, the purpose is to determine whether the signal intensity coming from the analyzed sample is positive or negative. Under this situation, it could be considered a reduction of the problem to have a signal different than the surrounding background. There are many examples in the literature where this approach has been followed and the analyses compare these signals with those coming from negative controls. Even in these simple approaches, it is necessary to keep a statistical control for the comparisons, and the experiments must be run under a multivariate environment.

Table 4.3 List of microarray data analysis software

Resource name	Purpose	Availability
Bioconductor	Bioinformatics repository with scripts developed for R language. Microarray data analysis, storage, and annotation tools for gene expression, genotyping, and other approaches. Desktop software	www.bioconductor.org
BRB array tools	Software developed in R with an MS Excel interface. There are two different add-ins, one focused on gene expression analysis and another based on CGH analysis. Desktop software	http://linus.nci.nih.gov/ BRB-ArrayTools.html
Cluster/cluster 3.0	Latest version of M. Eisen's popular Cluster software. This software is a developed clustering analysis program and provides some preprocessing options. Desktop software	http://bonsai.hgc. jp/~mdehoon/software/ cluster/ http://rana.lbl.gov/ EisenSoftware.htm
dChip	Software developed for either gene expression or SNP data analysis using Affymetrix microarray platforms	https://sites.google.com/ site/dchipsoft/
GenePattern	It is a genomic data analysis platform that includes tools for microarray data analysis; either gene expression or SNP analysis. It can be used either as desktop software or as a web tool	http://www.broadinstitute. org/cancer/software/ genepattern/
J-Express	Gene expression analysis software that includes preprocessing tools and functional analysis tools	http://jexpress.bioinfo.no
BABELOMICS	Web-based tool for gene expression and functional analysis. It includes a complete suite for data analysis from the preprocessing steps to the final functional analyses	www.babelomics.org
SAM/PAM	SAM (Significance Analysis of Microarrays) is one of the most common pieces of software used for class comparison in microarrays. PAM (Prediction Analysis of Microarray) is a program developed for class prediction	http://www-stat.stanford. edu/~tibs/SAM/index. html http://www-stat.stanford. edu/~tibs/PAM/index. html
MapleTree TreeView	Software developed for the visualization of results coming from the Cluster program based on Eisen's TreeView. It is possible to use it to represent results from hierarchical and other clustering algorithms	http://mapletree. sourceforge.net/ http://rana.lbl.gov/ EisenSoftware.htm

(continued)

Table 4.3 (continued)

Resource name	Purpose	Availability
TM4	Complete suite for microarray analysis comprised of four different software packages that can be found individually: Microarray Data Manages (MADAM), TIGR Spotfinder, Microarray Data Analysis System (MIDAS), and Multiexperiment Viewer (MeV)	http://www.tm4.org/
GeneSpring GX	Commercial/transcriptomics, genomics, metabolomics, and proteomics	Agilent http://www.agilent.com
Ingenuity pathway assist	Commercial/pathway analysis software	Ingenuity systems http://www.ingenuity.com
GeneMaths XT	Commercial/software for gene expression analysis	Applied maths
Spotfire –decision site for functional genomics	Commercial/data mining software for gene expression	Tibco http://spotfire.tibc.com
Nexus expression	Commercial/gene expression analysis	Biodiscovery http://www.biodiscovery.com
ArrayStar	Commercial/software for gene expression as well as sequence variation data across a group	DNA star http://www.dnastar.com
Partek genomics suite	Commercial/software for microarray and next-generation sequencing data analysis	Partek http://www.partek.com
Genedata expressionist	Commercial/comprehensive software for –omics data (transcriptomics, metabolomics, proteomics, etc.)	Genedata http://www.genedata.com
IBD	Commercial/software for gene expression analysis	Integromics http://www.integromics.com

The reality is more complex than that. In those "simple" approaches, discrimination done by multiple probes is necessary to ensure the statistical significance of that positive call considering all the remaining probes present in the microarray. In other experimental approaches where allele-specific oligonucleotide probes are used, either with hybridization or enzymatic reactions, base calling requires the use of specific analytical tools. In eukaryotic organisms when allele-specific probes are used, the success of the base genotype assignation methodologies relies on their capability to truly discern among three different situations: homozygous wild type/wild type, heterozygous wild type/mutant, or homozygous mutant/mutant. The use of different platforms implies the use of different approaches to perform the base calling. For example, Affymetrix developed its own application for base calling on its chips and

Table 4.4 List of some microarray statistical data analysis tools used for gene set enrichment analysis

Microarray statistical data analysis software freely available for academia/pathway analysis		
Resource name	Purpose	Availability
GSEA	Gene set enrichment analysis. Desktop	http://www.broadinstitute.org/gsea/index.jsp
DAVID	Gene set analysis. Online	http://david.abcc.ncifcrf.gov/
GSA	Gene set enrichment analysis. Desktop	http://www-stat.stanford.edu/~tibs/GSA/

Table 4.5 List of some tools designed for the analysis of genotyping microarrays. This list does not include proprietary tools that are available by the high-density microarray manufacturers for analysis of their arrays

Microarray software for genotyping		
Resource name	Purpose	Availability
Accutyping	Algorithms for data normalization Freely available for academia	http://www2.umdnj.edu/lilabweb/publications/AccuTyping.html
SNP chart	Analysis of APEX-based microarrays It is able to analyze data starting from images It is offered on a fee basis for academia	http://www.snpchart.ca
Genorama	Software developed specifically for APEX microarrays. Commercial	http://www.genorama.com
MACGT	Clustering-based genotyping tool Freely available for academia	http://www.hli.ubc.ca/research/PIs/MACGT.shtml
ResqMi	Designed for resequencing microarrays Freely available for academia	http://www-ps.informatik.uni-tuebingen.de/resqmi/
Bioconductor	Scripts developed for R language Microarray data analysis, storage, and annotation tools for gene expression, genotyping, and other approaches Desktop software	http://www.bioconductor.org

has been used as the basis for other tools. Despite the availability of this and other commercial platform-related specific tools, the scientific community has developed efforts to generate other new tools such as those available at the Bioconductor site. Table 4.5 contains a list of some software available for genotyping data analysis.

In the gene expression section, the fact that microarray data need to undergo a preprocessing step before any further analysis was brought up. In the case of genotyping analyses, these steps are also necessary, although the preprocessing activities are different. It is important to note that in many of these analyses, except for the more simplistic approaches, the normalization step is going to be necessary in the genotyping analyses as well, even if the normalization strategies are different from some of the ones developed for gene expression analyses. In some genotyping experiments, accepted requirements for global normalization strategies such as that

Table 4.6 Example of the relative allele signal (RAS) normalization method for genotyping data normalization

Normalization strategy	Explanation
Ratio-based: relative allele signal (RAS)	This normalization strategy is based on the comparison of ratios between the different probe pairs analyzed, wild type (wt) and mutant (mt). The normalized ratio (R) is generated for each spot calculating the ratio of each of the specific probes by the average sum of the intensities (I) of both probes. $R = I_{wt}/(I_{wt} + I_{mt})$

of having signals from a vast majority of spotted probes may not be fulfilled. Another important issue regarding normalization strategies in genotyping experiments is the variability due to the diversity of experimental approaches and the impact that the different strategies may have on the normalization requirements. It is not the same situation when several fluorophores are used, as in the APEX experiments, than when only one is used, as in the case of most of the hybridization-based approaches. The normalization step emphasizes intraslide normalization, since in comparative genome hybridization– (CGH) based approaches; interslide comparisons are not relevant beyond comparing whether a particular genotype is present or absent. These comparisons are done in absolute terms regardless of the intensity levels measured in the different experiments (Table 4.6).

When a single probe or color is used, some of the available analytical methods are based on the comparison of the obtained signals to those coming from negative controls or other controls. In many algorithms, the analytical methods developed for base calling are similar to those developed and used in gene expression analyses. Most of those approaches are based on the similarities between the problem of base calling and the class prediction approaches. In base calling, the classes are defined as the right and positive base assignations. Depending on the experimental approaches, it could be the perfect match compared to the mismatch or the differences between the normalized intensities among the different channels of known positive controls in multichannel experiments. Therefore, in these base-calling/class prediction approaches, many classes are defined as possible allelic variants that are analyzed in the experiments. Under this approach, it is possible to find a set of different tools using different statistical approaches such as linear discriminant analysis (LDA) or Bayesian networks. Another important set of tools used for base calling is based on the use of different variants of clustering algorithms.

4.4   Microarray Data Management and Standards

An important aspect related to every "-omics" approach is the management of vast amounts of information generated by these experiments. Microarrays pertain to one of these "-omics" technologies, and management is a key aspect for the success of the experiments.

Information management in microarray experimentation has evolved and undergone important changes during the development of the technologies and the methodologies utilized. When microarrays were originally developed, there was a lack of standards. Different platforms led to different processing schemas and data formats. Due to this wide variety of platforms, formats, and ways of analyzing and publishing data, the launch of different standardization projects was fostered by public and private consortia. After some divergent approaches in the development of these standardization initiatives, they finally merged and crystallized into the development of a set of recommendations about the parameters and information required for the annotation of microarray experiments. These would be transformed into the first and most successful standards for microarrays to date, the *Minimal Information About a Microarray Experiment* (MIAME). The MIAME standard was developed by the MGED group, now the Functional Genomics Data Society (FGED-Society), and published in 2001 (Brazma et al. 2001). It specifies all the information to be provided with any microarray experiment to enable the interpretation of its results in an unambiguous way, which facilitates any reproduction of the experiment for validation. The development of this standard was quickly adopted by major publishers requiring the submission of microarray data in a way compliant with this standard as well as encouraging the development of new public and private repositories for the storage of these experiments. Since MIAME is an exhaustive annotation standard for all microarray experiments, the major topics covered by the standard are

1. The experimental design. This defines the factors related to how the experiment is carried out.
2. Microarray annotation. This topic includes those aspects related to microarray manufacturing processes, the probes (their sequences, the sequences used for their design, and their physicochemical properties), and the probe distribution (the grid used in the microarray).
3. Annotation of experimental factors and samples. This important aspect of the MIAME standard describes all the procedures that affect the samples (such as nucleic acid extraction, purification, or labeling) and those experimental factors included in the study (for example, different treatments or different strains).
4. Protocol annotation. All the protocols followed during the studies should be included within the information provided with the microarray experiments. The protocols should include not only the "wet lab" protocols used but also those used for the data and information processing.
5. Raw data. The concept of raw data included in the MIAME standard refers to the quantification files generated after image analysis. These files should allow for the reproduction of any data analysis carried out and explore other ways of using different data preprocessing steps.
6. Processed data. In MIAME, the final processed data are included. These data are the preprocessed data and should include only the normalized and transformed remaining data points after filtering.

An important point related to the development of the MIAME standard is the usage of well-established and -defined controlled vocabularies and ontologies for the description and annotation of any of the processes involved in microarray experimentation. This recommendation is based on the fact that the use of such resources reduces the variability in the annotation and improves the quality and understanding of the annotated procedures.

The success of the MIAME initiative developed by the FGED society prompted the development of similar initiatives in other fields. It led to other standards in "-omics"-based approaches such as MINSEQE for its use in high-throughput nucleotide sequencing experiments.

Another area containing available standards for microarrays deals with the representation of microarray gene expression data and information data exchange. This other standard is called MAGE (MicroArray and Gene Expression). It is another standard developed by the FGED society and is based on the development of a set of tools that enables and facilitates the data exchange process. For this purpose, an XML-based language has been proposed (MAGE-ML) (Spellman et al. 2002). The recently developed MAGE-TAB spreadsheet format is the recommended option nowadays, since it is simpler than MAGE-ML, which is only used by those laboratories with a strong background in bioinformatics or additional support. The development of MAGE-TAB has finally spread the use of MAGE for data exchange without the need for knowing or understanding the details related to the XML-based format.

4.4.1 Microarray Data Repositories

As mentioned previously, during microarray experiments, huge amounts of information are generated and managed. The development of the standards and the requirement of publishers to provide information related to the experiments led to the development of public data repositories. In the early years, the information associated with experiments was presented in websites as singe files or sets of files. Later some groups or institutions started to develop their own storage facilities by saving the information and data that they considered most important. With the advent of the MIAME standard and the efforts of the FGED Society, the basis for the development of the major public repositories available nowadays were set. Despite the existence of these major public repositories, there are still other smaller resources focused more on everyday work in smaller laboratories or internal use in core facilities. These smaller and private data repositories should follow the MIAME standard and, if possible, the MAGE-TAB as well in order to simplify the transfer of information to the major public resources.

In an analogy to the previously cited world of nucleic acid sequences, the microarray world has followed some of the same paths, such as the requirement of depositing data in public repositories before publishing. There is also a similarity between microarray public repositories and databases and sequence databases. It is possible to find major public resources associated with the major sequence providers (EBI, NCBI, DDBJ), but there are also other participants involved that provide interesting resources, such as Stanford University. An extremely important difference between microarray resources and nucleotide databases is that there is not a complete data exchange policy among the nucleotide databases. Therefore, it is possible to find unique contents in each of these databases.

The Gene Expression Omnibus (GEO 2011) is the NCBI solution for microarray storage. The aim of this resource is to store abundant molecular species. This aim is broader than mere microarray storage. In this resource, it is possible to find data coming from other platforms different from microarrays, such as high-throughput sequencing data and other microarray data not focused on gene expression, such as protein microarrays and SNP microarrays.

The system has been designed to fulfill all the MIAME requirements and encourage the submitters to submit their data in a MIAME compliant way by means of a MIAME checklist-like submission procedure. In those cases where the submitters do not follow the MIAME standard, they should provide the raw data associated with each submission. Once the experimenters have submitted their data, it can be either public or kept private while a manuscript is under review. The reviewers may have access to this information via a temporary accession number for the datasets. The data submitted to GEO do not necessarily have to be referred to or supported by a publication. In cases where the data have been kept private and the NCBI staff detects that the dataset has been published, there is an immediate release of the data to make the data freely accessible to the whole community. Data submitted to GEO are accessible and grouped into two different categories: the GEO series (GSE) and the GEO datasets (GDS). GEO series are the original records related to a study submitted by an experimenter, while the GEO datasets are based on the GSEs, which have been processed and curated by GEO staff. These represent a collection of data that are comparable biologically and statistically.

Once the data of interest have been found, it is possible to download the data for further analyses. The downloadable files are a series matrix from the GSE or the full SOFT files from the GDS, and both files can be used for further analysis using different software.

GEO includes some embedded tools for data analysis associated with the stored information. These tools allow the user to perform statistical analysis for the identification of specific gene profiles within the selected study. Available tools include unsupervised techniques (clustering and heatmaps for the visualization of clustering results) and supervised techniques (t-test for sample comparison). It is important to note that access to the analytical tools embedded in the repository is available only for the GDS and not for the GSE.

ArrayExpress (2011), hosted by the European EBI, is a repository for array-based experiments as well as other functional data coming from RNA-Seq data. (RNA-Seq data come from the use of next-generation sequencing in gene expression studies.) Organized in two major components, the goals of ArrayExpress are to serve the community as a repository supporting publications, provide access to data in a standardized format, and facilitate the sharing of technical platforms and protocols. One of the major characteristics of ArrayExpress is its priority of storing well-annotated data.

The two elements constituting the ArrayExpress (AE) system are the ArrayExpress experiment archive, the database containing all submitted data, and the Gene

Expression Atlas. The Gene Expression Atlas (2011) is a curated and semantically enriched database of data generated from the meta analysis of data stored at the AE experiment archive.

An interesting characteristic of AE is that they are following a policy of importation of experiments stored in the GEO. AE imports experiments based on the Affymetrix and Agilent microarray catalogues on a weekly basis. These experiments (both GDS and GSE) are evaluated, and if they pass the quality filtering process (no corrupted files) are reannotated and incorporated into AE. These experiments contain a specific AE accession number E-GEOD-XXXX, which identifies an experiment as coming from this import process. A similar process is carried out with the Stanford University microarray database. In the case of the experiments submitted to the Stanford Microarray database, the accession number is E-SMD-XXXX.

The ArrayExpress database was designed by several of the groups involved in the development of the MIAME standard. For this reason, the database has been MIAME-compliant since its origin. The focus in covering the MIAME standard is reflected in the use of a MIAME score for the available data. This MIAME score represents the coverage of the MIAME standard in five major categories (array designs, protocols, factors, processed data, and raw data). The score is used to indicate the quality of the annotation provided by the submitter in those different areas.

Similar to the procedures developed for data uploading to GEO, AE has two different approaches for data uploading. An online strategy based on the use of their tool MIAMExpress allows the upload of experiments with fewer than 50 hybridizations and with a single raw file per hybridization, and a MAGE-TAB–based procedure as a general uploading method.

Database querying in AE for experiment retrieval can be done in two different steps. First, a general search is done using an ArrayExpress accession number or using keywords in the search textbox. Once this first search has been done, it is possible to apply some filters related to the species, technologies, specific arrays, or the molecular assay. It is possible to build complex queries combining search terms or specifying the search fields. The database offers specific querying interfaces for platform design or protocol retrieval.

An important difference between AE and GEO is that GEO contains some embedded analysis tools, while AE used to have a direct link to export its data to the ExpressionProfiler tool. Since this tool was retired at the end of 2011, this link is no longer available. The data may be downloaded to be analyzed in other third party tools.

Following the same schema used in nucleotide databases, what was supposed to be the third microarray database available is the Japanese database maintained by the DDBJ Center for Information Biology gene EXpression database (CIBEX 2011). Although when it was designed, it was expected to be the third major resource, it did not accomplish that goal. This database contains limited data, mostly submitted by Japanese groups.

In the world of microarrays, the role of the third major database is played by the microarray database of Stanford University, SMD (2011). The importance of this database is based on the influence that Stanford University had in microarray development. Stanford University has been involved in the development of microarray technologies since the very beginning and has played a key role in the spread of the technologies and some of the analytical procedures for data analysis. SMD use is free for the faculty of Stanford University and can be used by external groups who pay a fee depending on the number of arrays loaded. The difference between this database and the other resources is that it can be used for data storage while experiments are running instead of being used only for final storage. The data of this database can also be found either on GEO, ArrayExpress, or both. SMD is a MIAME-compliant database.

While accessing SMD, it is possible to explore the data associated with the most recent publications shown on the home website. In querying the database it is possible to use three different strategies: experiment-based, dataset-based or gene-based. The first search strategy allows the user to access the advanced querying interface.

SMD contains embedded functionality for data analysis that allows the user to process the raw data in the database by utilizing different preprocessing strategies. It supports two unsupervised clustering methods for data analysis.

4.5   Microarray Quality Control and Assessment

The results of microarray-based applications have been under suspicion for a long time due to issues related to the different technologies in microarray manufacturing, variability in protocols and experimental approaches, and finally, their complex analysis and the need for mathematical calculations for data interpretation. The lack of other standards different from those developed for data storage and information exchange did not help to add confidence in the technology. For these reasons, many efforts have been done to assess the quality and reproducibility of experiments carried out using microarrays. The standardization initiatives were focused on different areas, some of them having already been discussed in the previous section that dealt with MIAME and MAGE-ML. A crucial initiative to ensure the quality of microarray experiments was the MicroArray Quality Control project (MAQC).

This project has been developed in two different phases, MAQC-I and MAQC-II, funded by the U.S. Food and Drug Administration (FDA). The aim of these projects was the evaluation of the reproducibility, comparability, and performance of the different microarray platforms and laboratory techniques. The results of the first MAQC effort focusing on technological assessment were published in 2005 and 2006. They showed a high internal reproducibility among the different microarray platforms included in the study (the most commonly used ones, including commercial and some in-house printed microarrays) (Shi et al. 2006; Patterson et al. 2006; Chen et al. 2007). Another important aspect of these results was the comparability and agreement of the microarray results between the platforms and compared with other techniques, such as quantitative PCR. An important detail regarding these studies is that they were based on the comparability of gene expression studies. Genotyping microarrays have been generally validated by sequencing.

Microarray data analysis is still a challenging area for specialists in bioinformatics. Different methods have been developed and published for microarray data analysis. In the past, there were some attempts to standardize these analyses. All of these initiatives failed to establish an approved standard, but the closest result was a European Union–funded project (EMERALD) focused on promoting quality metrics for quality control of the experiments.

The second phase of the MAQC project (MAQC-II) was focused on the analysis of class prediction studies. These are some of the most relevant studies in health applications. The results of this second study were published in 2010 and showed that the biological endpoint is the most important aspect related to the performance of the algorithms used and analyzed (MAQC 2010; Luo et al. 2010). In this study, a set of five observations and three recommendations focused on the improvement of the procedures used for model building of the predictors were proposed.

The data availability that comes with the requirement of publishing results in public databases has enabled the possibility of carrying out analyses without doing any experiments by simply recovering and reanalyzing the data. Another option that is currently available is the possibility of running meta-analysis or complement studies by recovering data from other similar studies.

4.6   Probe Design Example Using PremierBiosoft's Array Designer 4.2

1. Open the software and open a new project File-> New Project.
2. Once the project is created, it is necessary to define the kind of probes that are going to be designed. This software allows the user to design probes under the following conditions: standard array design, tiling array design, whole-genome array design, and resequencing array design. There are differences in the conditions and the kind of probes that can be designed under these standard protocols. In this example, we will follow two examples in the design of the standard arrays and a resequencing array.
3. Once the design strategy is chosen, for example, standard array design, the next step is loading the sequences into the software. This process can be done in different ways, either opening the sequences from a local file or uploading them directly from GenBank using their accession numbers. Once the sequences are uploaded for a particular design strategy, they cannot be used in a different strategy and must be loaded again with a newly chosen probe design strategy.
4. Once the sequences are uploaded, it is possible to define the reaction conditions. These parameters will be used for the calculation of the thermodynamic properties of the probes (Fig. 4.6).
5. The next step consists of designing the probes and defining the parameters for the desired probes. The parameters that can be customized by the user are the length range of the probes (allowing the design of short or long probes), the T_m range, and the location of the probes. The advanced parameters tab opens a new

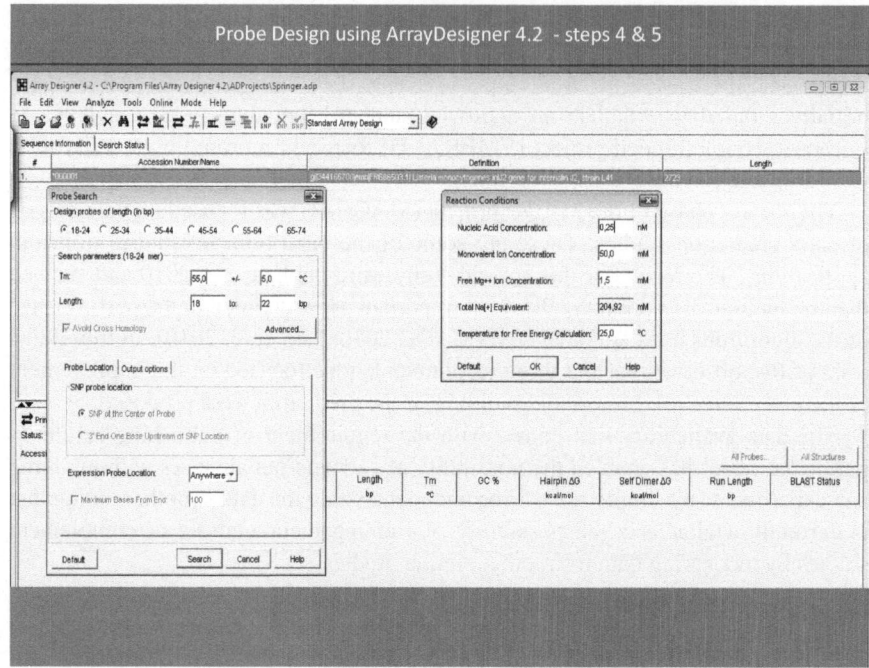

Fig. 4.6 Steps 4 & 5 in probe design using ArrrayDesigner 4.2 Software

dialog box containing some other specificity parameters. The Output tab allows the number of probes designed, the number of alternatives for each design, and the polarity of the design to be defined, whether it is based on the sense or anti-sense strand of the target sequence (Fig. 4.6).

6. In the case of designing probes for the analysis of SNPs, the SNPs should be introduced using the "Add SNP" button. Once the SNP is added, it appears in the lower half of the screen. If there are several SNPs in the same sequence, they must be introduced separately (Fig. 4.7). Another option is to introduce the same sequence several times as if it was different sequences and then set each SNP in each of these new sequences. The software designs probes for only one of the possible SNPs introduced in the sequence, so introducing the same sequence several times and setting each SNP in each copy is particularly useful to avoid the selection of each SNP one by one.

   At this stage, it is also possible to define the parameters used in the design of probes for SNP analysis. The two possibilities are the design of hybridization probes where the mismatch is located at the central position of the probe or a primer extension probe with the SNP located at the 3′ end one base upstream of the SNP location (Fig. 4.7).

7. Once the parameters for the probe design are set, the software performs the design and shows the number of probes created for each of the analyzed sequences. The resulting probes are listed in both parts of the screen (Fig. 4.7). In the upper half of the screen are the sequences and the probes in the second

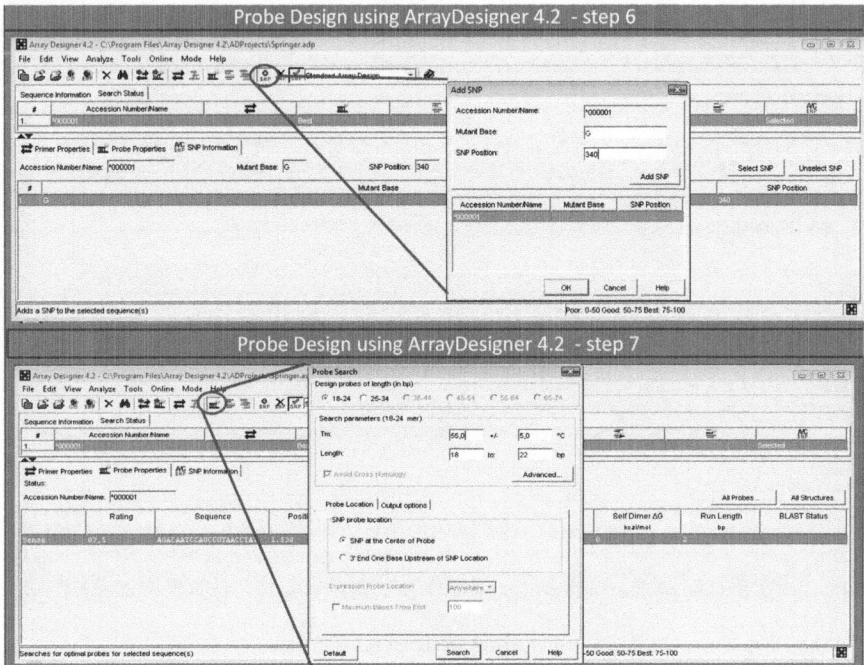

Fig. 4.7 Steps 6 & 7 in probe design using ArrrayDesigner 4.2 Software

column (probe design column). If a probe has been designed, the quality based on an internal scoring formula of the best probe designed is indicated. For SNP probe designs, the last column provides information about the SNP for which the probe has been designed. Appearing in the lower half of the screen are the detailed characteristics of the probe designed based on the sequence selected in the upper half of the screen.

8. Once the probes are designed, the next step is the control of their specificity by means of a BLAST search. ArrayDesigner 4.2 offers several search possibilities based on whether the BLAST is run on the same computer that the software is installed, run by using a local database defining a particular server, or run by using NCBI's BLAST services. With NCBI's BLAST services, it is also possible to define different search types based on the sequence datasets used to build the BLAST database. The offered NCBI-based searches are based on the nr database BLAST, and genome-based BLAST searches are focused on the Human Genome BLAST, Eukaryotic Genome BLAST, or Microbial Genome BLAST. In the case of the selection of any of the genome-based BLAST searches, it is necessary to use the "set genome" option to select the genome of interest as well as the database in case there are several available databases related to any particular genome.

Pressing the "Analyze" button offers the BLAST option. In this screen, it is possible to select the kind of designed element for which the BLAST search will be performed. For this example, the element is the designed probe (Fig. 4.8).

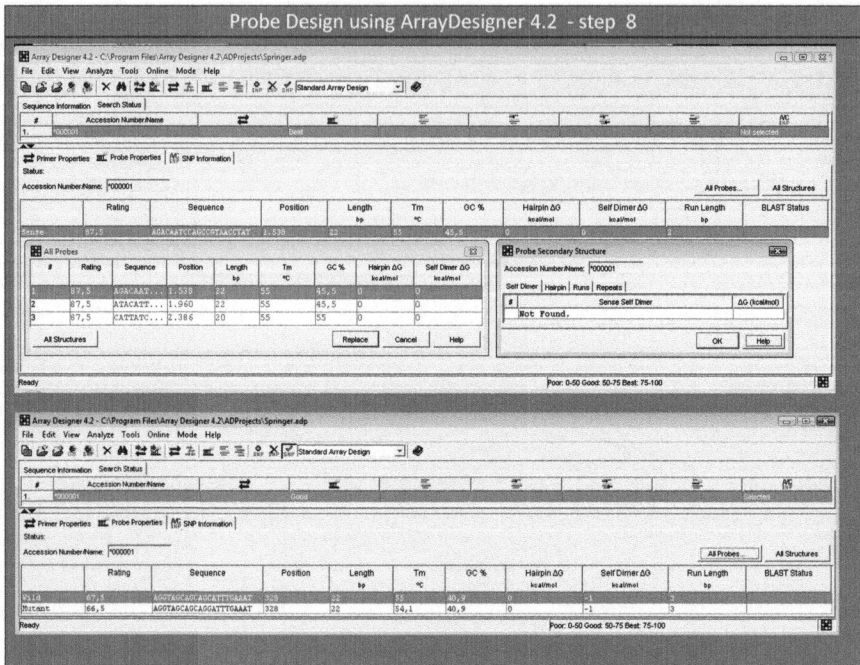

Fig. 4.8 Step 8 in probe design using ArrrayDesigner 4.2 Software

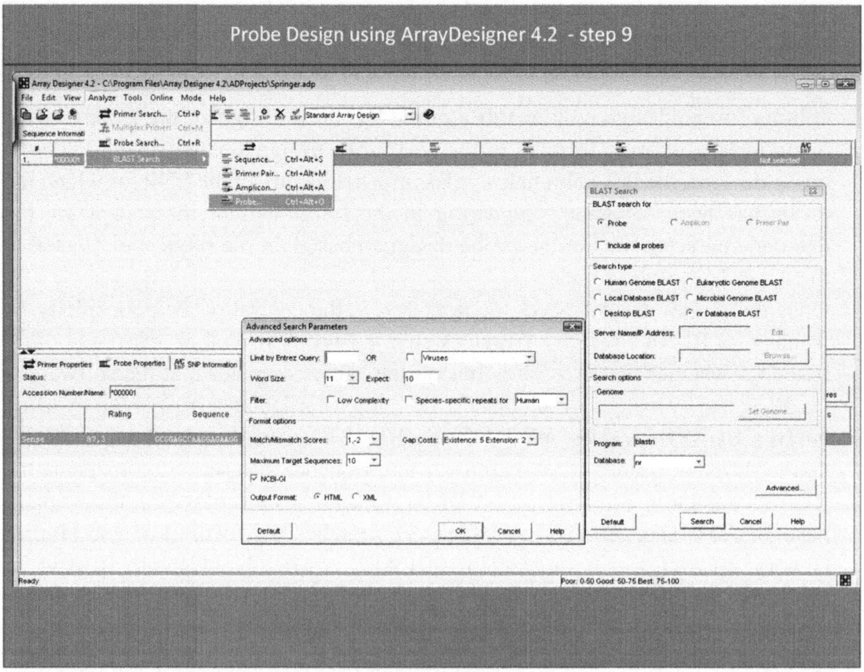

Fig. 4.9 Step 9 in probe design using ArrrayDesigner 4.2 Software

Pressing the "Advanced" button opens a new window offering different BLAST parameters, such as options for the searches (Fig. 4.8).

9. Once the probes have been designed, it is possible to save the project or to export the designed probes. In both cases, each option can be found under the "File" tab. The probes are exported as a spreadsheet, and before exporting them, one can decide the characteristics of the probes to be included in the spreadsheet (Fig. 4.9).

References

ArrayExpress (2011) http://www.ebi.ac.uk/arrayexpress. Accessed 28 November 2011.

Autio R, Kilpinen S, Saarela M, Kallioniemi O, Hautaniemi S, Astola J (2009) Comparison of Affymetrix data normalization methods using 6,926 experiments across five array generations. BMC Bioinformatics10:S24.

Benson DA, Karsch-Mizrachi I, Lipman DJ, Ostell J, Sayers EW (2011) GenBank Nucleic Acids Res 39:D32-D37.

Biotic (2011) http://biotic.isciii.es/paginas/diseño%20de%20sondas_2.htm. Accessed 28 November 2011.

Bolstad BM, Irizarry RA, Astrand M, Speed TP (2003) A comparison of normalization methods for high density oligonucleotide array data based on variance and bias. Bioinformatics 19:185–193.

Brazma A, Hingamp P, Quackenbush J, Sherlock G, Spellman P, Stoeckert C, Aach J, Ansorge W, Ball CA, Causton HC, Gaasterland T, Glenisson P, Holstege FC, Kim IF, Markowitz V, Matese JC, Parkinson H, Robinson A, Sarkans U, Schulze-Kremer S, Stewart J, Taylor R, Vilo J, Vingron M (2001) Minimum information about a microarray experiment (MIAME) toward standards for microarray data. Nat Genet 29:363–371.

Chen JJ, Hsueh HM, Delongchamp RR, Lin CJ, Tsai CA (2007) Reproducibility of microarray data: a further analysis of microarray quality control (MAQC) data. BMC Bioinformatics 8:412.

Chou CC, Chen CH, Lee TT, Peck K (2004) Optimization of probe length and the number of probes per gene for optimal microarray analysis of gene expression. Nucleic Acids Res 32:e99.

CIBEX (2011) DDBJ Center for Information Biology gene EXpression database http://cibex.nig.ac.jp/index.jsp. Accessed 28 November 2011.

Cochrane G, Akhtar R, Bonfield J, Bower L, Demiralp F, Faruque N (2009) Petabyte-scale innovations at the European Nucleotide Archive. Nucleic Acids Res 37:D19–D25.

Cooper KL, Goering RV (2003) Development of a universal probe for electronic microarray and its application in characterization of the *Staphylococcus aureus* polC gene. J Mol Diagn 5:28–33.

Davidsen T, Beck E, Ganapathy A, Montgomery R, Zafar N, Yang Q (2010) The comprehensive microbial resource. Nucleic Acids Res 38:D340–D345.

Ding Y, Wilkins D (2004) The effect of normalization on microarray data analysis. DNA Cell Biol 23:635–642.

Do JH, Choi DK (2006) Normalization of microarray data: single-labeled and dual-labeled arrays. Mol Cells 22:254–261.

Dotsch A, Pommerenke C, Bredenbruch F, Geffers R, Haussler S (2009) Evaluation of a microarray-hybridization based method applicable for discovery of single nucleotide polymorphisms (SNPs) in the *Pseudomonas aeruginosa* genome. BMC Genomics10:29.

Ember SW, Schulze H, Ross AJ, Luby J, Khondoker M, Giraud G, Terry JG, Ciani I, Tlili C, Crain J, Walton AJ, Mount AR, Ghazal P, Bachmann TT, Campbell CJ (2011) Fast DNA and protein microarray tests for the diagnosis of hepatitis C virus infection on a single platform. Anal Bioanal Chem 401:2549–2559.

Favis R, Gerry NP, Cheng YW, Barany F (2005) Applications of the universal DNA microarray in molecular medicine. Methods Mol Med 114:25–58.

GEO (2011) http://www.ncbi.nlm.nig.gov/geo. Accessed 28 November 2011.

Gerry NP, Witowski NE, Day J, Hammer RP, Barany G, Barany F (1999) Universal DNA microarray method for multiplex detection of low abundance point mutations. J Mol Biol 292:251–262.

Gillespie JJ, Wattam AR, Cammer SA, Gabbard JL, Shukla MP, Dalay O (2011) PATRIC: the Comprehensive Bacterial Bioinformatics Resource with a Focus on Human Pathogenic Species. Infect Immun 79:4286–4298.

Girigoswami A, Jung C, Mun HY, Park HG (2008) PCR-free mutation detection of BRCA1 on a zip-code microarray using ligase chain reaction. J Biochem Biophys Methods 70:897–902.

Herrero J, Diaz-Uriarte R, Dopazo J (2003) Gene expression data preprocessing. Bioinformatics 19:655–656.

Jarvinen AK, Laakso S, Piiparinen P, Aittakorpi A, Lindfors M, Huopaniemi L, Piiparinen H, Mäki M (2009) Rapid identification of bacterial pathogens using a PCR- and microarray-based assay. BMC Microbiol 9:161.

Kakinuma K, Fukushima M, Kawaguchi R (2003) Detection and identification of *Escherichia coli*, *Shigella*, and *Salmonella* by microarrays using the gyrB gene. Biotechnol Bioeng 83:721–728.

Kaminuma E, Kosuge T, Kodama Y, Aono H, Mashima J, Gojobori T (2011) DDBJ progress report. Nucleic Acids Res 39:D22–D27.

Kane MD, Jatkoe TA, Stumpf CR, Lu J, Thomas JD, Madore SJ (2000) Assessment of the sensitivity and specificity of oligonucleotide (50mer) microarrays. Nucleic Acids Res 28:4552–4557.

Kanehisa M, Goto S (2000) KEGG: Kyoto encyclopedia of genes and genomes. Nucleic Acids Res 28:27–30.

Kostic T, Stessl B, Wagner M, Sessitsch A, Bodrossy L (2010) Microbial diagnostic microarray for food- and water-borne pathogens. Microb Biotechnol 3:444–454.

Kostic T, Weilharter A, Rubino S, Delogu G, Uzzau S, Rudi K, Sessitsch A, Bodrossy L (2007) A microbial diagnostic microarray technique for the sensitive detection and identification of pathogenic bacteria in a background of nonpathogens. Anal Biochem 360:244–254.

Luo J, Schumacher M, Scherer A, Sanoudou D, Megherbi D, Davison T Shi T, Tong W, Shi L, Hong H, Zhao C, Elloumi F, Shi W, Thomas R, Lin S, Tillinghast G, Liu G, Zhou Y, Herman D, Li Y, Deng Y, Fang H, Bushel P, Woods M, Zhang J (2010) A comparison of batch effect removal methods for enhancement of prediction performance using MAQC-II microarray gene expression data. Pharmacogenomics J 10:278–291.

MAQC-II (2010) analyze that! Nat Biotechnol 28:761.

Park T, Yi SG, Kang SH, Lee S, Lee YS, Simon R (2003) Evaluation of normalization methods for microarray data. BMC Bioinformatics 4:33.

Patterson TA, Lobenhofer EK, Fulmer-Smentek SB, Collins PJ, Chu TM, Bao W, Fang H, Kawasaki ES, Hager J, Tikhonova IR, Walker SJ, Zhang L, Hurban P, de Longueville F, Fuscoe JC, Tong W, Shi L, Wolfinger RD (2006) Performance comparison of one-color and two-color platforms within the MicroArray Quality Control (MAQC) project. Nat Biotechnol 24: 1140–1150.

Rotter A, Hren M, Baebler S, Blejec A, Gruden K (2008) Finding differentially expressed genes in two-channel DNA microarray datasets: how to increase reliability of data preprocessing. OMICS 12:171–182.

Shi L, Reid LH, Jones WD, Shippy R, Warrington JA, Baker SC, MAQC Consortium (2006) The MicroArray Quality Control (MAQC) project shows inter- and intraplatform reproducibility of gene expression measurements. Nat Biotechnol 24:1151–1161.

SMD (2011) Database of Stanford University. http://smd.stanford.edu/. Accessed 28 November 2011.

Spellman PT, Miller M, Stewart J, Troup C, Sarkans U, Chervitz S, et al. (2002) Design and implementation of microarray gene expression markup language (MAGE-ML). Genome Biol 23:3.

Steinhoff C, Vingron M (2006) Normalization and quantification of differential expression in gene expression microarrays. Brief Bioinform 7:166–177.

The Gene expression Atlas (2011) http://www.ebi.ac.uk/gxa/. Accessed 28 November 2011.

Chapter 5
Applications of DNA Microarrays to Study Bacterial Foodborne Pathogens

Abstract DNA microarrays bring practical benefits to the field of food safety in regard to pathogen detection and characterization. This chapter is a review of how DNA microarray technology has been applied to the identification of pathogens, assessment of their diversity, analysis of their responses to different forms of stress, and studies on how specific metabolic pathways might contribute to the growth and survival of the pathogens in a range of niches, such as the food processing environment or a human host. Taken together, the studies reviewed in this chapter give significant information about how microarray platforms pertain to one of two main types. One type is more focused on genomic analysis for foodborne bacterial detection and characterization, while the other is more oriented toward the systemic research of pathogens in order to discover their molecular mechanisms of action. This is achieved by performing genome-wide screens in real time at affordable prices and with reliable outputs. DNA microarrays are simple to use and have a high potential for application in studies involving tracing pathogens through the food supply chain, particularly when rapid information about detection is important. Finally, microarray technology could also allow specific and sensitive testing for the assessment of virulence and antibiotic resistance of foodborne bacteria with the objectives of improving prevention, detection, and follow-up.

Keywords DNA microarrays • Gene expression profiling • Detecting foodborne pathogens • Genotyping of foodborne bacteria • Resequencing microarrays • Virulence of foodborne bacteria • Antimicrobial resistance of foodborne bacteria

5.1 DNA Microarrays to Study Bacterial Foodborne Diseases

DNA microarrays can be used to conduct two fundamental types of experiments: gene expression profiling and genotyping. In gene expression profiling experiments, the probes are designed to interrogate RNA. However for genotyping experiments,

G. López-Campos et al., *Microarray Detection and Characterization of Bacterial Foodborne* 93
Pathogens, SpringerBriefs in Food, Health, and Nutrition, DOI 10.1007/978-1-4614-3250-0_5,
© Guillermo Lopez-Campos, Joaquin V. Martinez-Suarez, Mónica Aguado-Urda,
Victoria Lopez Alonso 2012

the targets are DNA that has been extracted from biological samples and the probes are designed to survey the sequence variations in or among the samples Although it is a simple experimental approach, DNA microarray technology provides the basis for profound studies of comparative and functional genomics in foodborne pathogens, as will be further elaborated in the following sections of this chapter.

5.2   Microarrays to Obtain Gene Expression Profilings to Study Foodborne Pathogens

Gene expression analysis with microarrays was one of the first applications that successfully detected thousands of labeled target molecules in parallel. These quantitative assays are extremely useful in monitoring global changes in gene expression under various growth conditions, chemical treatments, or other stress-causing effects. Furthermore, data describing microbial gene expression in cultured cells and in experimental animals by microarrays are helping to resolve the mechanisms underlying pathogenesis. Overall, DNA expression arrays provide a very powerful tool for understanding how foodborne bacteria cope and survive in novel and constantly changing environments.

With advances in bioinformatics and microarray manufacturing, the quality of microarray data has improved dramatically, while the challenges associated with microarray cost and data analysis and interpretation are being resolved. The complete genome sequences for diverse bacterial strains belonging to the most important foodborne pathogenic species are available in the Genomes OnLine Database (GOLD) (Liolios et al. 2010) and in the Comprehensive Microbial Resource (CMR) (Schoolnik 2002). However, there is a lack of information on genes not present in these reference strains, and it is not possible to determine gene positions when making comparisons between reference strains and target strains. These are two examples of issues that still need to be worked out in microarray technology.

Initially, all the different microarray platforms are able to provide comparable results. A high level of agreement in gene expression microarray assays has been obtained even when employing different microarray types or facilities, different methods of purifying genomic DNA, or different wash buffers. However, less agreement between microarray results has been observed when using different hybridization buffers, indicating the importance of this parameter when transferring standard microarray assays between laboratories. As with any assay, replicate samples are critical for statistical analysis, because the fluorescent readout of hybridization intensities can vary between different laser scanners and reproducibility can vary greatly between different laboratories. The recognition of biases and other artifacts by individual labs and organizations, such as the MicroArray Quality Control (MAQC) consortium, has led to the development of quality control standards that operate to ensure the usefulness of a well-performed microarray experiment.

A new generation of high-density tiling arrays incorporates probes throughout the full bacterial genome to use for gene expression experiments. The Pathogen

Functional Genomics Resource Center (PFGRC 2011) distributes 70-mer glass slide DNA microarrays for different pathogens. The Bacterial Microarray Group at St George's (BμG@S 2011), established through the Wellcome Trust's Functional Genomics Initiative, makes DNA microarray technology available to the microbial pathogen research community on the basis of scientific collaboration. The only Affymetrix microarray available for pathogen gene expression is the GeneChip® *E. coli* Antisense Genome Array, which is used for examining the expression of all known *E. coli* genes. A variety of companies, such as Roche NimbleGen and Agilent Technologies, provide platforms for a more flexible approach to gene expression array design.

In order to support the analysis of information generated by gene expression arrays, the Food and Drug Administration (FDA) has developed free software, ArrayTrack™, which has recently been upgraded to manage and analyze genetic profiling data related to bacterial foodborne pathogens (Xu et al. 2010).

Disease pathogenesis reflects the complex interactions among host, pathogen, and environmental factors that influence both host and pathogen responses. Research on bacterial pathogenesis is a rapidly growing field as additional bacterial genome sequences become more and more available and techniques for conducting high-throughput analysis like microarrays are refined. At the transcriptome level, microarrays have frequently been used to measure mRNA abundance under certain conditions, enabling, for example, global surveys of several groups of associated virulence factors, antibiotic resistance, and stress-response genes in response to diverse environments and hosts.

Matching the expression profile produced by a particular stress or external condition with an archive of pathway-specific profiles should greatly improve the decision-making process, as in the case of treatment with antimicrobial compounds. Moreover, through the use of specific inhibitors whose sites of action are known, it can be possible to identify signature profiles that are specific for a particular pathway.

A main objective is to identify gene expression differences between strains that occupy different host niches or vary in some other subtle manner. Table 5.1 lists some studies done on the expression profiles of the most important foodborne bacteria types, *Escherichia coli* O157:H7, *Salmonella enterica* spp., *Listeria monocytogenes*, *Campylobacter jejuni*, and *Yersinia enterocolitica*.

5.2.1 Identification of Expression Profiles in Pathogens Associated with Virulence and Pathways of Host–Pathogen Interactions

To understand the bacterial response that occurs in the course of host–pathogen interactions, it is necessary to identify the genes required for infection and in vivo survival of the pathogen along with the genes that are differentially regulated in the host. Ideally, studies of this kind would compare expression profiles of bacteria from infected tissues with profiles of bacteria cultured under standardized in vitro growth conditions. Unfortunately, this goal has not been achieved without the use of

Table 5.1 Some examples of gene expression experiments with DNA arrays to study virulence factors and adaptation to environment changes of foodborne bacteria

Foodborne bacteria	Gene expression study	Reference(s)
Escherichia coli O157:H7	*lrp* Mutant versus wild type to study genes associated with nutrient limitation and osmotic stress	Tani et al. 2002
	dam Mutant versus wild type to study genes involved in regulation of the metabolic and respiratory pathways	Oshima et al. 2002
	Induction of stress associated transcripts	Dahan et al. 2004
	Baro-resistant versus baro-sensitive strains	Malone et al. 2006
	Mutants and different O_2 concentrations	Overton et al. 2006
	Response to multiple stresses of the food environment	Bergholz et al. 2009
	Growth in raw ground beef extract versus tryptic soy broth	Fratamico et al. 2011
	Survival at 15°C in sterile soil and natural water	Duffitt et al. 2011
Salmonella enterica	*pnp* Mutant versus wild type to study how PNPase affects genes involved in intracellular growth in mice	Clements et al. 2002
	csrA Mutant versus wild type to study this gene as a global regulator	Lawhon et al. 2003
	Repression of genes by probiotic effect	Keersmaecker et al. 2005
	Systemic infection in mice	Lawley et al. 2006
	Intramacrophage gene expression study	Faucher et al. 2006
	lpp Mutant versus wild type to study differences in flagellar and virulence gene expression	Fadl et al. 2006
	Response to chlorine oxidative stress	Wang et al. 2010
Yersinia enterocolitica	Temporal global changes during temperature transition	Motin et al. 2004
	Global gene regulation of virulence genes and FliA involved in flagellar synthesis	Horne and Pruss 2006
	Regulation of stress-response genes of phage-shock protein	Seo et al. 2007
	Virulence in mice	Rogers et al. 2007
	Transcriptional profile inside J774.1 macrophage-like cells	Fukuto et al. 2010
Listeria monocytogenes	σ^B Mutants of lineages I, II, IIIA, and IIIB to determine the contributions of σ^B to stress response and virulence	Oliver et al. 2010
	Expression of stress response and virulence genes	Rantsiou et al. 2012
	Growth on a ready-to-eat meat matrix to understand adaptation and survival ability	Bae et al. 2011
	Pressure-tolerant mutant strain to understand survival under high-hydrostatic-pressure (HHP) treatment	Liu et al. 2011
	Strain resistant to the bacteriocin sakacin	Tessema et al. 2011

(continued)

Table 5.1   (continued)

Foodborne bacteria	Gene expression study	Reference(s)
Campylobacter jejuni	Mechanisms involved in adaptation to the host (rabbit)	Flint et al. 2010
	Growth phase–dependent activation of the DccRS regulon	Wösten et al. 2010
	Growth in microaerobic atmospheres	John et al. 2011

amplification methods, because the number of organisms that can be obtained from infected tissues is normally small, and host cell RNA is vastly more abundant than bacterial RNA.

Bacterial virulence gene expression is affected by the growth phase, nutrient levels, temperature, oxygen, carbon source, pH, osmolarity, divalent cations, and iron concentrations, as has been shown in several studies. Taking these considerations into account, most microarray studies of pathogenic bacteria have employed growth conditions that are achievable in vitro to simulate host environments with respect to one or more of the mentioned parameters.

Microarray analysis has been used to study possible mechanisms involved in the adaptation of foodborne bacteria to the host by detailing the upregulation of the genes involved in ribosomal protein synthesis and modification, heat-shock response, and primary adaptation to the host environment. The regulation of virulence factors is expressed in many cases in a coordinate manner as shown in the flagellar gene expression of *E. coli* or *S. enterica* serovar Typhimurium, which consists of more than 40 genes. Early genes such as *flhCD* and *flgM* code for master regulator proteins, which direct the expression of intermediate genes (flagellar hook and basal body) and late flagellar genes (structural genes). Microarray experiments revealed 287 putative targets of FlhC, representing a variety of cellular functions such as central metabolism, cell division, biofilm formation, and pathogenicity. This has provided a valuable insight into the complex regulatory network of the pathogen that enables its survival under various environmental conditions (Sule et al. 2011).

Host genomic information has also yielded insights into bacterial virulence through the transcriptional profiling of host responses to infection and the identification of host proteins required for bacterial pathogenicity by using the knockdown of host gene product expression during infection (Sturdevant et al. 2010).

5.2.2   Studies of Pathogen Responses to Environmental Changes and Food Preservation Methods

The application of microarrays to different food models and environments has allowed the transcriptional profiling analysis of pathogens and the ability to identify complex pathways and strategies that a pathogenic bacterium utilizes in response to environmental stress or to food preservation methods (e.g., heating, chilling, acidity, alkalinity, etc.). Surprisingly, few studies focusing on issues directly relevant to food safety have been published, dealing with, for example, treatments with sanitizers, formation of biofilms, or growth on different food processing surfaces. Global approaches using microarrays to study foodborne bacterial response to diverse preservation methods will provide important keys to better control contamination in the food industry.

Microarray analyses of bacterial gene expression give information on genes that are involved in pathogenesis and virulence, stress response, and antibiotic resistance, all of which could be investigated as potential targets for food safety interventions. For example, in attempts to further understand the continuous problems that *E. coli* O157:H7 causes, different studies have been carried out with microarrays, obtaining the global gene regulation of this bacterium while growing on varying food matrices. These studies have provided insight into how this pathogen is able to survive, how exposure to food influences subsequent transmission and virulence, and how the bacteria are able to develop a different phenotype during transport through the environment (Bergholz et al. 2009; Duffitt et al. 2011).

The expression profiles of bacteria depend highly on the diverse stress conditions to which they are exposed:

- High-temperature conditions. Heat-shock proteins (HSPs) function as molecular chaperones and play an important role in protecting the cell from damage imposed by heat and other stresses. Whole-genome expression profiles of cells exposed to 48°C were examined using DNA microarrays to investigate the response in *L. monocytogenes* and *C. jejuni* (van der Veen et al. 2007). These studies reported that numerous differentially expressed genes that have roles in the cell division machinery or cell wall synthesis were downregulated, in line with the observation that heat shock results in cell elongation and the prevention of cell division.
- Mechanical stress. Mechanical stress injury produced from treatment with high hydrostatic pressure (HPP) was studied in *L. monocytogenes* following treatments with 400 and 600 MPa for 5 min. Results showed that HPP increased the expression of genes associated with DNA repair mechanisms, transcription and translation of protein complexes, the septal ring, the general protein translocase system, flagella assemblage, chemotaxis, and lipid and peptidoglycan biosynthetic pathways (Bowman et al. 2008).
- Acidic or alkaline environments. It is generally accepted that low pH induces a global stress response and regulates the expression of virulence genes in several bacterial species. Adaptation to acidic environments also has been studied with microarrays in the case of *E. coli*. When growing with acetate, the bacteria significantly increases its acid resistance and exhibits the upregulation of at least 26 genes, 6 of which were previously known to play a role in protecting the intracellular pH during acidic conditions of growth (Arnold et al. 2001). Investigations into resistance or susceptibility to certain types of environmental stresses like chlorine led to the identification of 209 core genes of *S. enterica* associated

with Fe–S cluster assembly, cysteine biosynthesis, stress response, ribosome formation, biofilm formation, and energy metabolism, all of which were differentially expressed (Wang et al. 2010).

- Regulation by bacterial density. Quorum sensing is a mechanism by which bacteria sense and respond to their own population density by means of signaling molecules, called autoinducers. Quorum sensing affects the expression of many cellular processes, including that of virulence genes, and has been studied in *E. coli* with microarray technology (Ren et al. 2004).

5.2.3 Studies of Pathogen Responses and Resistance to Antimicrobials

There is a great deal of complexity in gene expression responses of bacteria to antimicrobial treatment. There are four kinds of changes in gene expression signatures that are produced: (1) changes in expression of bacterial genes that are altered as a direct consequence of the effect on a target by an antimicrobial; (2) expression changes of genes indirectly affected by the antimicrobial effect on the primary target, including genes related to general stress responses, metabolic changes, and resistance mechanisms; (3) expression responses related to the secondary effect in co-regulated genes; and (4) changes in unrelated genes. Databases of gene expression signatures using microarrays for a range of antimicrobials and regulatory interactions are available for several microorganisms (e.g., RegulonDB for *E. coli* and DBTBS for *Bacillus subtilis*). The DBTBS (Hutter et al. 2004) predicts the mechanism of action for a series of antimicrobials.

5.3 Microarrays to Study Variations in DNA Sequence

5.3.1 Detection of Foodborne Bacteria

The highly miniaturized and parallel nature of DNA microarray technology permits the simultaneous assessment of the presence of many foodborne bacterial pathogens. Additionally, microarray technology can be applied to examine differences in gene content between taxonomically related strains that vary with respect to subtype and pathogenic phenotype.

A major challenge of microarrays is the parallel detection and identification of pathogens of low abundance within a complex microbial community. Prior to enabling the detection of a low number of bacterial cells in complex samples by DNA microarray assays, broad-range PCR amplification and subsequent fragmentation is necessary in many cases. This poses limits on the assay and also may cause a degree of bias. The critical step for the PCR approach is the selection of the right gene so that the sequence is conserved sufficiently to be able to identify several genera or species and, at the same time, has enough polymorphisms present so that species identification is possible. Most scientists have based microarray probe selection for bacterial detection on the variable regions of 16S or 23S rRNA genes to be used for the identification of foodborne bacterial pathogens by using universal PCR primers for 16S or 23S rRNA or rDNA (Wang et al. 2007).

On the other hand, other studies have detected bacterial species by employing bacterial species-specific genes, including housekeeping genes, virulence factors, and antibiotic resistance determinants, instead of rDNA sequences. Specific nucleotide sequences unique to the microorganism of interest have to be identified during the design of the microarray by performing multiple alignments of nucleotide sequences derived from all microbial species. The specificity and sensitivity of the candidate probes have to be first confirmed by comparison with sequences in public databases by BLAST analyses and then reconfirmed by hybridization studies. The design of the selected probes must avoid cross-hybridization, have uniformity of annealing temperatures (or GC content), and in general possess multiple probes for each sequence. Microarrays for detection have only positive and negative control spots to ensure that threshold limits for positive responses are maintained and that nonspecific adsorption does not create a false-negative signal. The sensitivity of microarray assay for the detection of target foodborne bacteria has been reported in the range of 10^2–10^4 colony-forming units (CFU) (Kostić et al. 2011; Sule et al. 2011). It is important to consider that to be able to make precise sensitivity comparisons, standardization of DNA extraction protocols and the use of internal standards and interlaboratory comparisons are necessary.

There are many examples of applications of microarrays for foodborne bacterial detection, as shown in Table 5.2. For the detection of foodborne pathogens, high-density tiling arrays with perfect match/mismatch probes are not widely available, but low- or medium-density arrays or suspension bead arrays are easily implemented due to their flexibility. Some new formats like the ArrayTube (Clondiag GmbH, Jena, Germany) is also promising and could potentially be developed for detection assays (Batchelor et al. 2008).

The US FDA has developed a microarray for the simultaneous detection of several foodborne pathogens and their virulence factors, including *Listeria* spp., *Campylobacter* spp., *Staphylococcus aureus* enterotoxin genes, and *Clostridium perfringens* toxin genes (Sergeev et al. 2004). This microarray has three elements that increase confidence in the detection system: redundancy of genes, redundancy of oligonucleotide probes for a specific gene, and quality control probes to monitor array spotting and target DNA hybridization. Microarrays are important tools for establishing monitoring programs to follow Enterohemorrhagic *E. coli* (EHEC) contamination in animals and foodstuffs, as part of the European Food Safety Authority program (EFSA 2009), which requires both the detection of a range of virulence factors and serotype determination.

The use of a multipathogen detection microarray to evaluate standard culture-based methods in microbiological food analysis has been demonstrated for different strains of *L. monocytogenes*, *Salmonella*, *Campylobacter*, and *Y. enterocolitica* that

Table 5.2 Representative studies of applications of microarrays for detection and genotyping of foodborne pathogens

Study	Foodborne pathogen(s)/microarray probes	Reference(s)
Multipathogen detection	*Salmonella* spp., *Vibrio* spp., and *Shigella dysenteriae*/16S rDNA	Eom et al. 2007; Hwang and Cha 2010
	Escherihia coli, Shigella spp., *Salmonella* spp., *Aeromonas hydrophila, Vibrio cholerae, Mycobacterium avium, Mycobacterium tuberculosis, Helicobacter pylori, Proteus mirabilis, Yersinia enterocolitica,* and *Campylobacter jejuni/gyrB* gene	Järvinen et al. 2009
	Staphylococcus aureus, Vibrio parahaemolyticus, Listeria monocytogenes, Salmonella, Cronobacter spp., *Shigella, Escherichia coli* O157:H7, and *Campylobacter jejuni*/species-specific genes	Jin et al. 2009
	Escherichia coli O157:H7, *Salmonella enterica, Listeria monocytogenes,* and *Campylobacter jejuni*/species-specific genes	Suo et al. 2010
	Vibrio spp., *Escherichia coli, Salmonella* spp., *Shigella* spp., and *Yersinia enterocolitica*/species-specific genes	Kim et al. 2010
	Acinetobacter baumannii, Enterococcus faecalis, Enterococcus faecium, Haemophilus influenzae, Klebsiella pneumoniae, Listeria monocytogenes, Neisseria meningitidis, Staphylococcus aureus, Staphylococcus epidermidis, Streptococcus agalactiae, Streptococcus pneumoniae, and *Streptococcus pyogenes*/genes of topoisomerases and hypervariable regions	Kostić et al. 2011
	Staphylococcus aureus, Salmonella spp., and *Bacillus cereus*/nonsequenced genomic DNA probes	Lee et al. 2011
Genotyping by detection of virulence and/or antimicrobial resistance genes	*Salmonella, Shigella,* and *Escherichia coli/ eaeA, slt-I, slt-II, fliC, rfbE,* and *ipaH* genes	Chizhikov et al. 2001
	Campylobacter jejuni/genes of biosynthesis of flagella, lipopolysaccharides, and capsule	Dorrell et al. 2001
	Listeria spp./*iap, hly, inlB, plcA, plcB,* and *clpE* genes	Volokhov et al. 2002
	Escherichia coli/antimicrobial resistance and virulence genes	Dobrindt et al. 2003; Korczak et al. 2005; Bonnet et al. 2009

(continued)

Table 5.2 (continued)

Study	Foodborne pathogen(s)/microarray probes	Reference(s)
	Escherichia coli/17 tetracycline resistance genes, β-lactamase gene	Call et al. 2003
	Campylobacter spp./23S rDNA, *fur, glyA, cdtABC, ceuB-C,* and *fliY* genes	Keramas et al. 2003; Volokhov et al. 2003
	Listeria monocytogenes/mixed-genome	Borucki et al. 2003
	Escherichia coli O157/virulence genes	Wu et al. 2003; Garrido et al. 2006
	Listeria spp., *Campylobacter* spp., *Staphylococcus aureus*, and *Clostridium perfringens*/virulence genes	Sergeev et al. 2004
	Listeria monocytogenes/genes of *Listeria monocytogenes* EGDe, *Listeria innocua* CLIP 11262, and *Listeria monocytogenes* CLIP 80459	Doumith et al. 2004
	Salmonella spp./*invA* and *spvB* genes	Porwollik et al. 2004; Ahn and Walt 2005; Peterson et al. 2010
	Vibrio spp./*vvh, viuB, ompU, toxR, tcpI,* and *hlyA* genes	Panicker et al. 2004; Vora et al. 2005; Chen et al. 2011
	Staphylococcus spp., *Streptococcus* spp., *Clostridium perfringens,* and *Bacillus anthracis*/antimicrobial resistance genes	Perreten et al. 2005
	Salmonella spp./antimicrobial resistance and virulence genes	Chen et al. 2005; Van Hoek et al. 2005; Batchelor et al. 2008
	Shigella and *Escherichia coli*/O-serotype specific genes	Li et al. 2006
	Bacillus cereus/enterotoxins, phospholipases, and exotoxins genes	Chandler et al. 2006
	51 Distantly related bacteria, including Gram-negative and Gram-positive isolates/antimicrobial resistance genes	Frye et al. 2006
	Yersinia enterocolitica/16S rRNA and virulence genes	Myers et al. 2006
	Lactococcus garvieae/antibiotic resistance genes	Walther et al. 2008
	Salmonella enterica/antimicrobial resistance genes	Zou et al. 2009; Anjum et al. 2011
	Gram-positive and Gram-negative bacteria/ antimicrobial resistance genes	Garneau et al. 2010
	Salmonella enterica and *Escherichia coli*/ *Inc A/C* genes and *Inc* H1 plasmids	Lindsey et al. 2010
	Salmonella enterica/virulence genes	Hauser et al. 2010
	Lactobacillus rhamnosus/antimicrobial resistance genes	Korhonen et al. 2010

(continued)

Table 5.2 (continued)

Study	Foodborne pathogen(s)/microarray probes	Reference(s)
Genotyping by CGH	*Escherichia coli* strain K-12 MG1655/genome	Dobrindt et al. 2003
	Salmonella enterica serovar Typhimurium strain SL1344/genome	Chan et al. 2003
	Salmonella enterica strain LT2/genome	Alvarez et al. 2003; Reen et al. 2005
	Campylobacter jejuni/genome	Leonard et al. 2003; Taboada et al. 2004
	Listeria monocytogenes/genomes of six serotypes	Zhang et al. 2003; Borucki et al. 2004
	Yersinia pestis/genome	Zhou et al. 2004
	Salmonella spp./genome	Bae et al. 2005
	Plasmids of *Salmonella enterica* and *Escherichia coli*/genomes	Call et al. 2006
	Yersinia enterocolitica/genome	Thomson et al. 2006
	Escherichia coli, Shigella/genomes	Jackson et al. 2011
Genotyping by detection of SNPs	*Escherichia coli*/blaTEM-1and blaTEM-37 genes	Grimm et al. 2004
	Campylobacter spp./23S rDNA genes	Mamelli et al. 2005
	Escherichia coli O157:H7/pO157 plasmid	Zhang et al. 2006
	Escherichia coli/gyrA gene	Yu et al. 2007

were inoculated in food matrices of epidemiological relevance (e.g., meat, eggs, and cheese) (Suo et al. 2010). However, having microarray technology in the food industry will highly depend on automation. Studies should be carried out to validate automated methods for extraction, PCR amplification, labeling, and quality controls of bacterial DNA from different food matrices (Liu et al. 2004).

5.3.2   Genotyping of Foodborne Bacteria

Although the use of microarray technology in detecting and characterizing food-borne pathogens is still in its infancy compared to the use of this technology in other areas like drug discovery, toxicology, and clinical diagnosis, a properly designed genotyping microarray can provide strain discrimination within a particular patho-type, identify genetic elements responsible for virulence and antimicrobial resis-tance, elucidate the genomic evolution of a diverse pathogen genus, and carry out epidemiological investigations of an outbreak of foodborne illness. Microarray analysis found that isolates that were epidemiologically grouped together by other methods of molecular typing (such as serotyping) showed genetic variability, which led to the use of a genovar classification, as in the case of strains of the same serovar in *Salmonella* (Porwollik et al. 2004; Peterson et al. 2010) or isolates of *L. monocytogenes* (Borucki et al. 2003; Call et al. 2003) obtained from epidemiologically and geographically similar sources. The FDA ECSG array is an Affymetrix microarray that assays all the genes found in *E. coli* and a related genus, *Shigella*, to identify and discriminate among closely related strains (ECSG 2009; Jackson et al. 2011).

Some DNA microarrays are based on previous knowledge of polymorphic loci obtained using classical typing techniques such as multilocus sequence typing (MLST). These also include highly variable regions that are known to be involved in horizontal gene transfer (Swiderek et al. 2005; Litrup et al. 2010).

From an epidemiological point of view, microarrays offer the opportunity to screen large numbers of bacterial isolates for the specific presence of a large number of antimicrobial resistance genes (determined previously by gene expression analysis), with the goal of obtaining data about the prevalence and diffusion of resistance determinants over time and geographic areas while considering the predicted spread of an outbreak. A prime example of the potential of DNA microarray technology to detect as many resistance genes as possible can be found in the microarray built through the use of the target gene sequences found in GenBank (National Center for Biotechnology Information, NCBI). This method is being used as a screening technique in studies of prevalence, epidemiology, and spread of resistance genes in a variety of bacteria, including *S. enterica* serovars, *E. coli, Campylobacter* spp., *Enterococcus* spp., methicillin-resistant *S. aureus*, *Listeria* spp., and *C. difficile* (Frye et al. 2010). Another microarray was developed previously to identify resistance genes common in Gram-positive and Gram-negative bacteria (Perreten et al. 2005; Garneau et al. 2010).

Examples of microarrays used to detect resistance caused by enzymatic inactivation of antimicrobials include studies of genes that code for enzymes that modify aminoglycoside antibiotics or degrade ß-lactam drugs such as penicillins and cephalosporins (Grimm et al. 2004) or dihydrofolate antimicrobials like sulfonamides (Hopkins et al. 2007). Other resistance mechanisms studied are permeability changes such as tetracycline resistance (Call et al. 2003) and target alteration such as resistance to macrolides and (fluoro) quinolone antimicrobials (Cassone et al. 2006). A specific serovar of *S. enterica,* which is resistant to multiple antibiotics, was investigated in a study by microarray hybridization, revealing genetic deletions in the glyoxylate pathway and in the second-phase flagellar gene (Garaizar et al. 2002).

Microarrays have also been utilized to infer interesting and sometimes unexpected information about virulence-related genes. These genes and their products affect the ability of a pathogen to infect or survive in its host, are required for expression of other virulence factors, or cause the host direct harm, such as with toxins (see Table 5.2 for examples). The GeneDisc® array (Bugarel et al. 2010) has been developed for the identification of 12 O-types and 7 H-types of Shiga toxin-producing *E. coli* (STEC), including the most clinically relevant enterohemorrhagic *E. coli* (EHEC) serotypes, and it was used by Germany's National Reference Laboratory to expedite the testing of food samples for the EHEC O104:H4 strain causing the 2011 outbreak of infection.

This combination of gene typing for virulence and antimicrobial resistance with DNA microarrays is a genomics-based approach to bacterial molecular epidemiology used to study the evolution and/or transfer of mobile antibiotic-resistance and virulence-associated genes (Vernet et al. 2004; Cassone et al. 2006).

5.3.2.1   Genotyping Using Comparative Genomic Hybridization (CGH) Microarrays

Comparisons using CGH microarray technology between a fully sequenced reference pathogen and an unsequenced one can provide valuable information about diversity and evolution of the pathogen or about the emergence of multidrug-resistant isolates. Genomic comparisons between pathogenic and nonpathogenic strains of the same species can be particularly informative.

One approach of CGH microarrays is using whole-genome microarrays that incorporate most of the genes on the chromosome of fully sequenced reference bacteria. For example, the Affymetrix GeneChip® *E. coli* Genome 2.0 Array, originally designed for transcriptome analysis, can be used also as a genotyping tool (Jackson et al. 2011). Another possibility is to use a shotgun DNA library that includes open reading frames (ORFs) from different strains of the same serotype or a collection of reference serotypes. The resulting hybridized microarray will reveal genes common to both strains and genes that are only present in the reference strain. The main advantage of microarrays is that thousands of DNA sequences can be analyzed concurrently in target isolates with rapidity and accuracy, and also the number of gene copies in a bacterial strain can be determined. However, CGH microarrays will not be able to detect genes present in the experimental strain but absent in the reference strain. Similarly, point mutations, including frame-shift mutations, deletions in homologous repetitive elements, or rearrangements of the genome that have not resulted in a gene deletion will not be detected. Several examples of CGH microarrays used to study foodborne bacteria are listed in Table 5.2.

5.3.2.2   Identification and Detection of Relevant Single-Nucleotide Polymorphisms (SNPs) for the Screening of Virulence Factors and Antimicrobial Resistance

SNPs can be detected in isolates of the same ribotype and having basically identical pulsed-field gel electrophoresis (PFGE) patterns. For example, investigation into *L. monocytogenes* isolates from a multistate outbreak of food poisoning in 1998 found mutations associated with different virulence factors (Zhang et al. 2003; Borucki et al. 2004), and STEC O157 strains with increased human virulence have been detected with this technique (Clawson et al. 2009). Bacteria possess a broad genetic diversity and can have a number of mutations in diverse gene families associated with antimicrobial resistance (e.g., *bla*, *aac*, *erm*, *mecA*, *van*, *gyr*, etc.). Some examples are SNPs in the gyrase gene *gyrA* in *E. coli*, which leads to resistance to nalidixic acid or fluoroquinolones (Yu et al. 2007), and mutations in the β-lactamase genes *bla*TEM-1 and *bla*TEM-37, which produce resistance to ampicillin and to third- and fourth-generation cephalosporins (Grimm et al. 2004).

There are several different microarray platforms to screen for the presence of SNPs. All the assays belonging to these platforms (listed in Table 5.3) are based on one of two main chemistry technologies to detect SNPs with microarrays: hybridization or enzymatic methods using either DNA polymerase or DNA ligase.

Table 5.3 Characteristics of some microarray platforms for SNP genotyping

Platform	Methodology	Features	Nos. of SNPs and samples	Example
Affymetrix	Allele hybridization	GeneChip Resequencing microarray	2,000,000 SNPs, 1 sample	Jackson et al. 2011
Printed arrays	Allele primer extension Allele hybridization Ligation	Specific design Four/single coloor	50,000 SNPs, 1 sample	Pullat and Metspallu 2008; Lu et al. 2009
Illumina	Allele primer extension	iSelect bead array Universal zip code	1,000,000 SNPs, 4, 12, and 24 samples	Fan et al. 2006
Sequenom	Allele primer extension	MassArray	100,000 SNPs, 96, and 384 samples	Lee et al. 2011

The method based on hybridization requires that the immobilized probes include all possible sequences at the sites of the SNP and a stretch of nucleotide sequence flanking the SNP on each side. The probes differ by a single mismatch at the central position to reduce nonspecific hybridizations. This method becomes feasible with the use of arrays that can carry up to 10^6 probes/cm², as with the Affymetrix GeneChip®.

SNP genotyping based on enzymatic selectivity can be performed using DNA polymerase or DNA ligase. In an assay consisting of single-base extensions, also called microarray minisequencing or Arrayed Primer Extension (APEX), the DNA polymerase uses four fluorescently labeled dideoxynucleotides (ddNTPs) to extend the detection primer by a single nucleotide at the position of the SNP.

SNP detection using DNA ligase requires two specific probe pairs. One probe containing the SNP being assayed is attached to the microarray with the 3′ end free. The second probe consists of the sequence just after the SNP position and is attached to a fluorescence reporter. The SNP becomes detectable only if the fluorescence probe and the target DNA are joined together by DNA ligase after hybridization of the sample in the microarray.

Both technologies can be utilized in combination with the generic Tag-array system, sometimes called an "array of arrays." Tag-array utilizes generic capture oligonucleotides ("cTags") immobilized on the microarray, enabling the basic non-SNP-specific microarray design format. The reactions are performed in solution using primers that carry Tag sequences complementary to one of the arrayed cTags, and the SNPs are determined when the primers hybridize to their corresponding cTags with known locations on the array. This method is particularly useful for the genotyping of bacteria. Using this approach, a microarray was developed to detect eight common foodborne pathogens, *S. aureus*, *V. parahaemolyticus*, *L. monocytogenes*, *Salmonella*, *Cronobacter* spp., *Shigella*, *E. coli* O157:H7, and *C.*

jejuni, with a sensitivity of 0.1 pg for genomic DNA and 5×10^2 CFU/mL for *Salmonella* Typhimurium cultures (Lu et al. 2009). The Tag-array system has also been used as a universal ligation-detection microarray, for example, in the Premi®Test *Salmonella* (PTS) and the *Salmonella* array (SA). These are rapid, easy, and commercially available DNA microarray platforms for the identification and typing of *Salmonella,* with a sensitivity up to 98.4% and a specificity of 99.98% (Wattiau et al. 2008; Koyuncu et al. 2011).

This approach has led to promising techniques such as the use of padlock probes. Padlock probes are linear oligonucleotides with ends complementary to the target SNP and a central stretch of random sequence that hybridizes with the array (Kurt et al. 2009).

5.3.2.3 Resequencing Microarrays

Resequencing microarrays use sets of probes with one perfectly matched and three mismatched probes per base for both strands of target genes. Identifications require a priori knowledge of a differential hybridization pattern that is determined empirically with control experiments. Even when control experiments are carried out, these characteristic hybridization patterns do not always occur with highly diverse pathogen targets obtained from food specimens. One interesting feature of resequencing microarrays is the possibility of detecting nucleic acids in a sample, even if their sequence diverges by up to 10–15% from those that are represented on the array. A resequencing microarray serves to monitor bacterial evolution comprehensively over short time periods. This capability is important because many microbial phenomena, such as the emergence of new pathogens and the acquisition of antibiotic-resistance factors, can occur over relatively short time scales.

The only resequencing array made by Affymetrix is the GeneChip® *E. coli* Genome 2.0. This array represents the genic and intergenic sequences of four sequenced strains of *E. coli* that represent a total of 10,208 probe sets. Each probe set contains approximately 22 oligonucleotide probes, 11 perfect match (PM) probes, and 11 mismatch probes (MM). Mismatch probes are identical to the perfect match probe with the exception of a single-nucleotide mismatch located at the 13th (middle) position of the oligonucleotide sequence. These mismatch probes are designed to allow for the approximation and correction of nonspecific hybridization signals. Probe design strategies such as this would be ideally suited for genotyping studies for two reasons: first, the probe redundancy for each genomic target sequence, and second, the excellent specificity afforded by the hybridization of short 25-mer probes.

The US Centers for Disease Control and Prevention (CDC) have developed a broad-range resequencing pathogen microarray (RPM) for the detection of 84 types of pathogens and 13 toxin genes of tropical and emerging infectious agents that includes the detection of most of the class A, B, and C select biothreat agents (Leski et al. 2009).

5.4 Microarrays to Study Protein–DNA Interactions in Bacteria

The identification of protein–DNA interactions in bacteria plays a crucial role in studying transcription, replication, recombination, and DNA repair. While gene expression studies are useful for studying global patterns of transcription, they are of limited use for defining the regulons of transcription factors, which need to be studied in conjunction with protein-DNA interaction techniques. Many transcription factors regulate the expression of other proteins involved in the complex bacterial transcriptional regulatory network, and the deletion of a transcription factor often has many secondary effects, for example, on bacteria virulence.

Chromatin immunoprecipitation (ChIP) is an experimental method to determine both the position and the strength of protein–DNA interactions in vivo. ChIP has been combined with microarrays to create the ChIP–chip technique (sometimes referred to as ChIP-onchip or ChIP^2). Briefly, ChIP involves the cross-linking of DNA and promoter proteins with formaldehyde, followed by lysis and fragmentation. Next, the protein of interest together with the cross-linked DNA is immunoprecipitated, the cross-linking is reversed with heat, and the DNA is purified and hybridized to a microarray to be compared with a genomic DNA control. Until recently, ChIP–chip had been applied very little to bacteria, which are ideally suited to methodologies involving microarrays because of their small genome sizes.

References

Ahn S, Walt DR (2005) Detection of *Salmonella* spp. using microsphere-based fiber-optic DNA microarrays. Anal Chem 77:5041–5047.

Alvarez J, Porwollik S, Laconcha I, Gisakis V, Vivanco AB, Gonzalez I, Echenagusia S, Zabala N, Blackmer F, McClelland M, Rementeria A, Garaizar J (2003) Detection of a *Salmonella enterica* serovar California strain spreading in Spanish feed mills and genetic characterization with DNA microarrays. Appl Environ Microbiol 69:7531–7534.

Anjum MF, Choudhary S, Morrison V, Snow LC, Mafura M, Slickers P, Ehricht R, Woodward MJ (2011) Identifying antimicrobial resistance genes of human clinical relevance within *Salmonella* isolated from food animals in Great Britain. J Antimicrob Chemother 66:550–559.

Arnold CN, McElhanon J, Lee A, Leonhart R, Siegele DA (2001) Global analysis of *Escherichia coli* gene expression during the acetate-induced acid tolerance response. J Bacteriol 18:2178–2186.

BµG@S (2011) Bacterial Microarray Group at St George's http://bugs.sgul.ac.uk/. Accessed 28 November 2011.

Bae D, Crowley MR, Wang C (2011) Transcriptome analysis of *Listeria monocytogenes* grown on a ready-to-eat meat matrix. J Food Prot 74:1104–1111.

Bae JW, Rhee SK, Nam YD, Park YH (2005) Generation of subspecies level-specific microbial diagnostic microarrays using genes amplified from subtractive suppression hybridization as microarray probes. Nucleic Acids Res 33:e113.

Batchelor M, Hopkins KL, Liebana E, Slickers P, Ehricht R, Mafura M, Aarestrup F, Mevius D, Clifton-Hadley FA, Woodward MJ, Davies RH, Threlfall EJ, Anjum MF (2008) Development of a miniaturised microarray-based assay for the rapid identification of antimicrobial resistance genes in Gram-negative bacteria. Int J Antimicrob Agents 31:440–451.

Bergholz TM, Vanaja SK, Whittam TS (2009) Gene expression induced in *Escherichia coli* O157:H7 upon exposure to model apple juice. Appl Environ Microbiol 75:3542–3553.

Bonnet C, Diarrassouba F, Brousseau R, Masson L, Topp E, Diarra MS (2009) Pathotype and antibiotic resistance gene distributions of *Escherichia coli* isolates from broiler chickens raised on antimicrobial-supplemented diets. Appl Environ Microbiol 75:6955–6962.

Borucki MK, Krug MJ, Muraoka WT, Call R (2003) Discrimination among *Listeria monocytogenes* isolates using a mixed genome DNA microarray. Vet Microbiol 4:351–362.

Borucki MK, Kim SH, Call DR, Smole SC, Pagotto F (2004) Selective discrimination of *Listeria monocytogenes* epidemic strains by a mixed-genome DNA microarray compared to discrimination by pulsed-field gel electrophoresis, ribotyping, and multilocus sequence typing. J Clin Microbiol 42:5270–5276.

Bowman JP, Bittencourt CR, Ross T (2008) Differential gene expression of *Listeria monocytogenes* during high hydrostatic pressure processing. Microbiology 154:462–475.

Bugarel M, Beutin L, Martin A, Gill A, Fach P (2010) Microarray for the identification of Shiga toxin-producing *Escherichia coli* (STEC) seropathotypes associated with Hemorrhagic Colitis and Hemolytic Uremic Syndrome in humans. Int J Food Microbiol 142:318–329.

Call DR, Bakko MK, Krug MJ, Roberts MC (2003) Identifying antimicrobial resistance genes with DNA microarrays. Antimicrob Agents Chemother 47:3290–3295.

Call DR, Kang MS, Daniels J, Besser TE (2006) Assessing genetic diversity in plasmids from *Escherichia coli* and *Salmonella enterica* using a mixed-plasmid microarray. J Appl Microbiol 100:15–28.

Cassone MM, D'Andrea F, Iannelli MR, Oggioni GM, Rossolini G, Pozzi G (2006) DNA microarrays for detection of macrolide resistance genes. Antimicrob Agents Chemother 50:2038–2041.

Chan K, Baker S, Kim CC, Detweiler CS, Dougan G, Falkow S (2003) Genomic comparison of *Salmonella enterica* serovars and *Salmonella bongori* by use of a *S. enterica* serovar Typhimurium DNA microarray. J Bacteriol 185:553–563.

Chandler DP, Alferov O, Chernov B, Daly DS, Golova J, Perov A, Protic M, Robison R, Schipma M, White A, Willse A (2006) Diagnostic oligonucleotide microarray fingerprinting of *Bacillus* isolates. J Clin Microbiol 44:244–250.

Chen S, Zhao S, McDermott PF, Schroeder CM, White DG, Meng J (2005) A DNA microarray for identification of virulence and antimicrobial resistance genes in *Salmonella* serovars and *Escherichia coli*. Mol Cell Probes 19:195–201.

Chen W, Yu S, Zhang C, Zhang J, Shi C, Hu Y, Suo B, Cao H, Shi X (2011) Development of a single base extension-tag microarray for the detection of pathogenic *Vibrio* species in seafood. Appl Microbiol Biotechnol 89:1979–1990.

Chizhikov V, Rasooly A, Chumakov K, Levy DD (2001) Microarray analysis of microbial virulence factors. Appl Environ Microbiol 67:3258–3263.

Clawson ML, Keen JE, Smith TP, Durso LM, McDaneld TG, Mandrell RE, Davis MA, Bono JL (2009) Phylogenetic classification of *Escherichia coli* O157:H7 strains of human and bovine origin using a novel set of nucleotide polymorphisms. Genome Biol 10:R56.

Clements MO, Eriksson S, Thompson A, Lucchini S, Hinton JCD, Normark S, Rhen M (2002) Polynucleotide phosphorylase is a global regulator of virulence and persistency in *Salmonella enterica*. Proc Natl Acad Sci 99:8784–8789.

Dahan S, Knutton S, Shaw RK, Crepin VF, Dougan G, Frankel G (2004) Transcriptome of enterohemorrhagic *Escherichia coli* O157 adhering to eukaryotic plasma membranes. Infect Immun 72:5452–5459.

Dobrindt U, Agerer F, Michaelis K, Janka A, Buchrieser C, Samuelson M, Svanborg C, Gottschalk G, Karch H, Hacker J (2003) Analysis of genome plasticity in pathogenic and commensal *Escherichia coli* isolates by use of DNA arrays. J Bacteriol 185:1831–1840.

Dorrell N, Mangan JA, Laing KG (2001) Whole genome comparison of *Campylobacter jejuni* human isolates using a low-cost microarray reveals extensive genetic diversity. Genome Res 11:1706–1715.

Doumith M, Cazalet C, Simoes N, Frangeul L, Jacquet C, Kunst F, Martin P, Cossart P, Glaser P, Buchrieser C (2004) New aspects regarding evolution and virulence of *Listeria monocytogenes* revealed by comparative genomics and DNA arrays. Infect Immun 72:1072–1083.

Duffitt AD, Reber RT, Whipple A, Chauret C (2011) Gene expression during survival of *Escherichia coli* O157:H7 in soil and water. Int J Food Microbiol pii:340506.

ECSG (2009) The Pathogen Functional Genomics Resource Center at the J. Craig Venter Institute (JCVI) announces a program to obtain the Affymetrix *E. coli* FDA-ECSG microarray. http://pfgrc.jcvi.org/index.php/home/full_news/current/2009_1_9.html. Accessed 28 November 2011.

EFSA (2009) Scientific report of EFSA: technical specifications for the monitoring and reporting of verotoxigenic *Escherichia coli* (VTEC) on animals and food. The EFSA Journal 7:1–43.

Eom HS, Hwang BH, Kim DH, Lee IB, Kim YH, Cha HJ (2007) Multiple detection of foodborne pathogenic bacteria using a novel 16S rDNA based oligonucleotide signature chip. Biosens Bioelectron 22:845–853.

Fadl AA, Galindo CL, Sha J, Klimpel GR, Popov VL, Chopra AK (2006) Global gene expression of a murein (Braun) lipoprotein mutant of *Salmonella enterica* Typhimurium by microarray analysis. Gene 374:121–127.

Fan JB, Gunderson KL, Bibikova M, Yeakley JM, Chen J, Wickham Garcia E, Lebruska LL, Laurent M, Shen R, Barker D (2006) Illumina universal bead arrays. Methods Enzymol 410:57–73.

Faucher SP, Porwollik S, Dozois CM, McClelland M, Daigle F (2006) Transcriptome of *Salmonella enterica* serovar Typhi within macrophages revealed through selective capture of transcribed sequences. Proc Natl Acad Sci USA 103:1906–1911.

Flint A, Butcher J, Clarke C, Marlow D, Stintzi A (2010) Use of a rabbit soft tissue chamber model to investigate *Campylobacter jejuni*-host interactions. Front Microbiol 1:126.

Fratamico PM, Wang S, Yan X, Zhang W, Li Y (2011) Differential gene expression of *E. coli* O157:H7 in ground beef extract compared to tryptic soy broth. J Food Sci 76:M79–M87.

Frye JG, Jesse T, Long F, Rondeau G, Porwollik S, McClelland M, Jackson CR, Englen M, Fedorka-Cray PJ (2006) DNA microarray detection of antimicrobial resistance genes in diverse bacteria. Int J Antimicrob Agents 27:138–151.

Frye JG, Lindsey RL, Rondeau G, Porwollik S, Long F, McClelland M, Jackson CR, Englen MD, Meinersmann RJ, Berrang ME (2010) Development of a DNA microarray to detect antimicrobial resistance genes identified in the National Center for Biotechnology Information Database. Microb Drug Resist 16:9–19.

Fukuto HS, Svetlanov A, Palmer LE, Karzai AW, Bliska JB (2010) Global gene expression profiling of *Yersinia pestis* replicating inside macrophages reveals the roles of a putative stress-induced operon in regulating type III secretion and intracellular cell division. Infect Immun 78:3700–3715.

Garaizar J, Porwollik S, Echeita A, Rementeria A, Herrera S, Wong RM-Y, Frye J, Usera MA, Mc Clelland M (2002) DNA microarray-based typing of an atypical monophasic *Salmonella enterica* serovar. J Clin Microbiol 40:2074–2078.

Garneau P, Labrecque O, Maynard C, Messier S, Masson L, Archambault M, Harel J (2010) Use of a bacterial antimicrobial resistance gene microarray for the identification of resistant *Staphylococcus aureus*. Zoonoses Public Health 57 Suppl 1:94–99

Garrido P, Blanco M, Moreno-Paz M, Briones C, Dahbi G, Blanco J, Blanco J, Parro V (2006) STECEPEC oligonucleotide microarray: a new tool for typing genetic variants of the LEE pathogenicity island of human and animal Shiga toxin producing *Escherichia coli* (STEC) and enteropathogenic *E. coli* (EPEC) strains. Clin Chem 52:192–201.

Grimm V, Ezaki S, Susa M, Knabbe C, Schmid RD, Bachmann TT (2004) Use of DNA microarrays for rapid genotyping of TEM beta-lactamases that confer resistance. J Clin Microbiol 42:3766–3774.

Hauser E, Tietze E, Helmuth R, Junker E, Blank K, Prager R, Rabsch W, Appel B, Fruth A, Malorny B (2010) Pork contaminated with *Salmonella enterica* serovar 4,[5],12:i:-, an emerging health risk for humans. Appl Environ Microbiol 76:4601–4610.

Hopkins KL, Batchelor MJ, Anjum M, Davies RH, Threlfall EJ (2007) Comparison of antimicrobial resistance genes in nontyphoidal salmonellae of serotypes Enteritidis, Hadar, and Virchow from humans and food-producing animals in England and Wales. Microb Drug Resist 13:281–288.

Horne SM, Pruss BM (2006) Global gene regulation in *Yersinia enterocolitica*: effect of FliA on the expression levels of flagellar and plasmid-encoded virulence genes. Arch Microbiol 185: 115–126.

Hutter B, Schaab C, Albrecht S, Borgmann M, Brunner NA, Freiberg C, Ziegelbauer K, Rock CO, Ivanov I, Loferer H (2004) Prediction of mechanisms of action of antibacterial compounds by gene expression profiling. Antimicrob Agents Chemother 48:2838–2844.

Hwang BH, Cha HJ (2010) Pattern-mapped multiple detection of 11 pathogenic bacteria using a 16S rDNA-based oligonucleotide microarray. Biotechnol Bioeng 106:183–192.

Jackson SA, Patel IR, Barnaba T, LeClerc JE, Cebula TA (2011) Investigating the global genomic diversity of *Escherichia coli* using a multi-genome DNA microarray platform with novel gene prediction strategies. BMC Genomics 12:349.

Järvinen AK, Laakso S, Piiparinen P, Aittakorpi A, Lindfors M, Huopaniemi L, Piiparinen H, Mäki M (2009) Rapid identification of bacterial pathogens using a PCR and microarray based assay. BMC Microbiol 9:161.

Jin SQ, Yin BC, Ye BC (2009) Multiplexed bead-based mesofluidic system for detection of food-borne pathogenic bacteria. Appl Environ Microbiol 75:6647–6654.

John A, Connerton PL, Cummings N, Connerton IF (2011) Profound differences in the transcriptome of *Campylobacter jejuni* grown in two different, widely used, microaerobic atmospheres. Res Microbiol 162:410–418.

Keersmaecker SC, Marchal K, Verhoeven TL, Engelen K, Vanderleyden J, Detweiler CS (2005) Microarray analysis and motif detection reveal new targets of the *Salmonella enterica* serovar Typhimurium HilA regulatory protein, including HilA itself. J Bacteriol 187: 4381–4391.

Keramas G, Bang DD, Lund M, Madsen M, Rasmussen SE, Bunkenborg H, Telleman P, Christensen CB (2003) Development of a sensitive DNA microarray suitable for rapid detection of *Campylobacter* spp. Mol Cell Probes 17:187–196.

Kim DH, Lee BK, Kim YD, Rhee SK, Kim YC (2010) Detection of representative enteropathogenic bacteria, *Vibrio* spp., pathogenic *Escherichia coli*, *Salmonella* spp., *Shigella* spp., and *Yersinia enterocolitica*, using a virulence factor gene-based oligonucleotide microarray. J Microbiol 48:682–688.

Korczak B, Frey J, Schrenzel J, Pluschke G, Pfister R, Ehricht R, Kuhnert P (2005) Use of diagnostic microarrays for determination of virulence gene patterns of *Escherichia coli* K1, a major cause of neonatal meningitis. J Clin Microbiol 43:1024–1031.

Korhonen JM, Van Hoek AH, Saarela M, Huys G, Tosi L, Mayrhofer S, Wright AV (2010) Antimicrobial susceptibility of *Lactobacillus rhamnosus*. Benef Microbes 1:75–80.

Kostić T, Stessl B, Wagner M, Sessitsch A (2011) Microarray analysis reveals the actual specificity of enrichment media used for food safety assessment. J Food Prot 74:1030–1034.

Koyuncu S, Andersson G, Vos P, Häggblom P (2011) DNA microarray for tracing Salmonella in the feed chain. Int J Food Microbiol 145:S18–S22.

Kurt K, Alderborn A, Nilsson M, Strommenger B, Witte W, Nübel U (2009) Multiplexed genotyping of methicillin-resistant *Staphylococcus aureus* isolates by use of padlock probes and tag microarrays . J Clin Microbiol 47:577–585.

Lawhon SD, Frye JG, Suyemoto M, Porwollik S, McClelland M, Altier C (2003) Global regulation by CsrA in *Salmonella* Typhimurium. Mol Microbiol 48:1633–1645.

Lawley TD, Chan K, Thompson LJ, Kim CC, Govoni GR, Monack DM (2006) Genome-wide screen for *Salmonella* genes required for long-term systemic infection of the mouse. PLoS Pathog 2:e11.

Lee JY, Kim BC, Chang KJ, Ahn JM, Ryu JH, Chang HI, Gu MB (2011) A substractively optimized DNA microarray using non-sequenced genomic probes for the detection of food-borne pathogens. Appl Biochem Biotechnol 164:183–193.

Leonard EE II, Takata T, Blaser MJ, Falkow S, Tompkins LS, Gaynor EC (2003) Use of an open-reading frame-specific *Campylobacter jejuni* DNA microarray as a new genotyping tool for studying epidemiologically related isolates. J Infect Dis 187:691–694.

Leski TA, Lin B, Malanoski AP, Wang Z, Long NC, Meador CE, Barrows B, Ibrahim S, Hardick JP, Aitichou M, Schnur JM, Tibbetts C, Stenger DA (2009) Testing and validation of high density resequencing microarray for broad range biothreat agents detection. PLoS One 11:e6569.

Li Y, Liu D, Cao B, Liu Y, Liu F, Guo X, Bastin DA, Feng L, Wang L (2006) Development of a serotype-specific DNA microarray for identification of some Shigella and pathogenic Escherichia coli strains. J Clin Microbiol 44:4376–83.

Lindsey RL, Frye JG, Fedorka-Cray PJ, Welch TJ, Meinersmann RJ (2010) An oligonucleotide microarray to characterize multidrug resistant plasmids. J Microbiol Methods 81:96–100.

Liolios K, Chen IM, Mavromatis K, Tavernarakis N, Hugenholtz P, Markowitz VM, Kyrpides NC (2010) The Genomes On Line Database (GOLD) in 2009: status of genomic and metagenomic projects and their associated metadata. Nucleic Acid Res 38:D346–D354.

Litrup E, Torpdahl M, Malorny B, Huehn S, Christensen H, Nielsen EM (2010) Association between phylogeny, virulence potential and serovars of Salmonella enterica. Infect Genet Evol 10:1132–1139.

Liu Y, Ream A, Joerger RD, Liu J, Wang Y (2011) Gene expression profiling of a pressure-tolerant Listeria monocytogenes Scott A ctsR deletion mutant. J Ind Microbiol Biotech 38:1523–1533.

Liu, RH, Yang J, Lenigk R, Bonanno J, Grodzinski P (2004) Self-contained, fully integrated bio-chip for sample preparation, polymerase chain reaction amplification, and DNA microarray detection. Anal Chem 76:1824–1831.

Li Y, Liu D, Cao B, Liu Y, Liu F, Guo X, Bastin DA, Feng L, Wang L (2006) Development of a serotype-specific DNA microarray for identification of some Shigella and pathogenic Escherichia coli strains. J Clin Microbiol 44:4376–83.

Lu C, Shi C, Zhang C, Chen J, Shi X (2009) Development of single base extension-tags microarray for the detection of food-borne pathogens. Sheng Wu Gong Cheng Xue Bao 25:554–559.

Malone AS, Chung YK, Yousef AE (2006) Genes of Escherichia coli O157:H7 that are involved in high pressure resistance. Appl Environ Microbiol 72:2661–2671.

Mamelli L, Prouzet-Mauleon V, Pages JM (2005) Molecular basis of macrolide resistance in Campylobacter: role of efflux pumps and target mutations. Antimicrob Chemother 56:491–497.

Motin VL, Georgescu AM, Fitch JP, Gu PP, Nelson DO, Mabery SL, Garnham JB, Sokhansanj BA, Ott LL, Coleman MA, Elliott JM, Kegelmeyer LM, Wyrobek AJ, Slezak TR, Brubaker RR, Garcia E (2004) Temporal global changes in gene expression during temperature transition in Yersinia pestis. J Bacteriol 186:6298–6305.

Myers KM, Gaba J, Al-Khaldi SF (2006) Molecular identification of Yersinia enterocolitica iso-lated from pasteurized whole milk using DNA microarray chip hybridization. Mol Cell Probes 20:71–80.

Oliver HF, Orsi RH, Wiedmann M, Boor KJ (2010) Listeria monocytogenes {sigma} B has a small core regulon and a conserved role in virulence but makes differential contributions to stress tolerance across a diverse collection of strains. Appl Environ Microbiol 76:4216–4232.

Oshima T, Wada C, Kawagoe Y, Ara T, Maeda M, Masuda Y, Hiraga S, Mori H (2002) Genome-wide analysis of deoxyadenosine methyltransferase-mediated control of gene expression in Escherichia coli. Mol Microbiol 45:673–695.

Overton TW, Griffiths L, Patel MD, Hobman JL, Penn CW, Cole JA, Constantinidou C (2006) Microarray analysis of gene regulation by oxygen, nitrate, nitrite, FNR, NarL, and NarP during anaerobic growth of Escherichia coli: new insights into microbial physiology. Biochem Soc Trans 34:104–107.

Panicker G, Call DR, Krug MJ, Bej AK (2004) Detection of pathogenic Vibrio spp. in shellfish by using multiplex PCR and DNA microarrays. Appl Environ Microbiol 70:7436–7444.

Perreten V, Vorlet-Fawer L, Slickers P, Ehricht R, Kuhnert P, Frey J (2005) Microarray-based detection of 90 antibiotic resistance genes of Gram-positive bacteria. J Clin Microbiol 43:2291–2302.

Peterson G, Gerdes B, Berges J, Nagaraja TG, Frye JG, Boyle DS, Narayanan S (2010) Development of microarray and multiplex polymerase chain reaction assays for identification of serovars and virulence genes in Salmonella enterica of human or animal origin. J Vet Diagn Invest 22:559–569.

PFGRC (2011) The Pathogen Functional Genomics Resource Center at the J. Craig Venter Institute (JCVI) http://pfgrc.jcvi.org/. Accessed 28 November 2011.

Porwollik S, Boyd EF, Choy C, Cheng P, Florea L, Proctor E, Mc Clelland M (2004) Characterization of *Salmonella enterica* subspecies I genovars by use of microarrays. J Bacteriol 186:5883–5896.

Pullat J, Metspalu A (2008) Arrayed primer extension reaction for genotyping on oligonucleotide microarray. Methods Mol Biol 444:161–167.

Rantsiou K, Greppi A, Garosi M, Acquadro A, Mataragas M, Cocolin L (2012) Strain dependent expression of stress response and virulence genes of *Listeria monocytogenes* in meat juices as determined by microarray. Int J Food Microbiol 152:116–122.

Reen FJ, Boyd EF, Porwollik S, Murphy BP, Gilroy D, Fanning S, Mc Clelland M (2005) Genomic comparison of *Salmonella enterica* serovar Dublin, Agona, and Typhimurium strains recently isolated from milk filters and bovine samples from Ireland, using a *Salmonella* microarray. Appl Environ Microbiol 71:1616–1625.

Ren D, Bedzyk LA, Ye RW, Thomas SM, Wood TK (2004) Stationary phase quorum-sensing signals affect autoinducer-2 and gene expression in *Escherichia coli*. Appl Environ Microbiol 70:2038–2043.

Rogers JV, Choi YW, Giannunzio LF, Sabourin PJ, Bornman DM, Blosser EG, Sabourin CL (2007) Transcriptional responses in spleens from mice exposed to *Yersinia pestis* CO92. Microb Pathog 43:67–77.

Schoolnik, GK (2002) Functional and comparative genomics of pathogenic bacteria. Curr Opin Microbiol 5:20–26.

Seo J, Savitzky DC, Ford E, Darwin AJ (2007) Global analysis of tolerance to secretin-induced stress in *Yersinia enterocolitica* suggests that the phage-shock-protein system may be a remarkably self-contained stress response. Mol Microbiol 65:714–727.

Sergeev N, Distler M, Courtney S, Al-Khaldi SF, Volokhov D, Chizhikov V, Rasooly A (2004) Multipathogen oligonucleotide microarray for environmental and biodefense applications. Biosens Bioelectron 20:684–698.

Sturdevant DE, Virtaneva K, Martens C, Bozinov D, Ogundare O, Castro N, Kanakabandi K, Beare PA, Omsland A, Carlson JH, Kennedy AD, Heinzen RA, Celli J, Greenberg DE, De Leo FR, Porcella SF (2010) Host-microbe interaction systems biology: lifecycle transcriptomics and comparative genomics. Future Microbiol 5:205–219.

Sule P, Horne SM, Logue CM, Prüss BM (2011) Regulation of cell division, biofilm formation, and virulence by FlhC in *Escherichia coli* O157:H7 grown on meat. Appl Environ Microbiol 77:3653–3662.

Suo B, He Y, Paoli G, Gehring A, Tu SI, Shi X (2010) Development of an oligonucleotide-based microarray to detect multiple foodborne pathogens. Mol Cell Probes 24:77–86.

Swiderek H, Claus H, Frosch M, Vogel U (2005) Evaluation of custom-made DNA microarrays for multilocus sequence typing of *Neisseria meningitidis*. Int J Med Microbiol 295:39–45.

Taboada EN, Acedillo RR, Carrillo CD, Findlay WA, Medeiros DT, Mykytczuk OL, Roberts MJ, Valencia CA, Farber JM, Nash JH (2004) Large-scale comparative genomics metaanalysis of *Campylobacter jejuni* isolates reveals low level of genome plasticity. J Clin Microbiol 42:4566–4576.

Tani TH, Khodursky A, Blumenthal RM, Brown PO, Matthews RG (2002) Adaptation to famine: a family of stationary-phase genes revealed by microarray analysis. Proc Natl Acad Sci USA 99:13471–13476.

Tessema GT, Møretrø T, Snipen L, Axelsson L, Naterstad K (2011) Global transcriptional analysis of spontaneous sakacin P-resistant mutant strains of *Listeria monocytogenes* during growth on different sugars. PLoS One 6:e16192.

Thomson NR, Howard S, Wren BW, Holden MT, Crossman L, Challis GL, Churcher C, Mungall K, Brooks K, Chillingworth T, Feltwell T, Abdellah Z, Hauser H, Jagels K, Maddison M, Moule S, Sanders M, Whitehead M (2006) The complete genome sequence and comparative genome analysis of the high pathogenicity *Yersinia enterocolitica* strain 8081. PLoS Genet 152:e206.

van der Veen S, Hain T, Wouters JA, Hossain H, de Vos WM, Abee T, Chakraborty T, Wells-Bennik MH (2007) The heat-shock response of *Listeria monocytogenes* comprises genes involved in heat shock, cell division, cell wall synthesis, and the SOS response. Microbiology 153:3593–3607.

Van Hoek AH, Scholtens IM, Cloeckaert A, Aarts UJ (2005) Detection of antibiotic resistance genes in different *Salmonella* serovars by oligonucleotide microarray analysis. J Microbiol Methods 62:13–23.

Vernet, G, Jay C, Rodriguez M, Troesch A (2004) Species differentiation and antibiotic susceptibility testing with DNA microarrays. J Appl Microbiol 96:59–68.

Volokhov D, Chizhikov V, Chumakov K, Rasooly A (2003) Microarray-based identification of thermophilic *Campylobacter jejuni, C. coli, C. lari*, and *C. upsaliensis*. J Clin Microbiol 41:4071–4080.

Volokhov DV, Rasooly A, Chumakov V, Chizhikov K (2002) Identification of *Listeria* species by microarray based assay. J Clin Microbiol 40:4720–4728.

Vora GJ, Meador CE, Bird MM, Bopp CA, Andreadis JD, Stenger DA (2005) Microarray-based detection of genetic heterogeneity, antimicrobial resistance, and the viable but nonculturable state in human pathogenic *Vibrio* spp. Proc Natl Acad Sci USA 102:19109–19114.

Walther C, Rossano A, Thomann A, Perreten V (2008) Antibiotic resistance in *Lactococcus* species from bovine milk: presence of a mutated multidrug transporter *mdt(A)* gene in susceptible *Lactococcus garvieae* strains. Vet Microbiol 131:348–357.

Wang XW, Zhang L, Jin LQ, Jin M, Shen ZQ, An S, Chao FH, Li JW (2007) Development and application of an oligonucleotide microarray for the detection of food-borne bacterial pathogens. Appl Microbiol Biotechnol 76:225–233.

Wang S, Phillippy AM, Deng K, Rui X, Li Z, Tortorello ML, Zhang W (2010) Transcriptomic responses of *Salmonella enterica* serovars Enteritidis and Typhimurium to chlorine-based oxidative stress. Appl Environ Microbiol 76:5013–5024.

Wattiau P, Weijers T, Andreoli P, Schliker C, Veken HV, Maas HM, Verbruggen AJ, Heck ME, Wannet WJ, Imberechts H, Vos P (2008) Evaluation of the Premi Test *Salmonella*, a commercial low-density DNA microarray system intended for routine identification and typing of *Salmonella enterica*. Int J Food Microbiol 123:293–298.

Wösten MM, van Dijk L, Parker CT, Guilhabert MR, van der Meer-Janssen YP, Wagenaar JA, van Putten JP (2010) Growth phase-dependent activation of the DccRS regulon of *Campylobacter jejuni*. J. Bacteriol 192:2729–2736.

Wu CF, Valdes JJ, Bentley WE, Sekowski JW (2003) DNA microarray for discrimination between pathogenic O157:H7 EDL933 and non-pathogenic *Escherichia coli* strains. Biosens Biochem 30:1–8.

Xu J, Kelly R, Fang H, Tong W (2010) ArrayTrack: a free FDA bioinformatics tool to support emerging biomedical research. Hum Genomics 4:428–434.

Yu X, Susa M, Weile J, Knabbe C, Schmid RD, Bachmann TT (2007) Rapid and sensitive detection of fluoroquinolone-resistant *Escherichia coli* from urine samples using a genotyping DNA microarray. Int J Med Microbiol 297:417–429.

Zhang C, Zhang M, Ju J, Nietfeldt J, Wise J, Terry PM, Olson M, Kachman SD, Wiedmann M, Samadpour M, Benson AK (2003) Genome diversification in phylogenetic lineages I and II of *Listeria monocytogenes*: identification of segments unique to lineage II populations. J Bacteriol 185:5573–5584.

Zhang W, Qi W, Albert TJ, Motiwala AS, Alland D, Hyytia-Trees EK, Ribot EM, Fields PI, Whittam TS, Swaminathan B (2006) Probing genomic diversity and evolution of *Escherichia coli* O157 by single nucleotide polymorphisms. Genome Res 16:757–767.

Zhou DY, Han E, Dai Y, Song D, Pei J, Zhai Z, Du J, Wang Z, Guo R, Yang R (2004) Defining the genome content of live plague vaccines by use of whole-genome DNA microarray. Vaccine 22:3367–3374.

Zou W, Frye JG, Chang CW, Liu J, Cerniglia CE, Nayak R (2009) Microarray analysis of antimicrobial resistance genes in *Salmonella enterica* from preharvest poultry environment. J Appl Microbiol 107:906–914.

Chapter 6
Future Trends

Abstract Detection technologies developed in the last 10 years have been directed to find more rapid, specific, and sensitive methods to detect multiple targets simultaneously. Much emphasis has been put on the desired characteristics of high sensitivity, high specificity, and cost-effectiveness, which are of great importance in routine high-volume food diagnostics. Developments in microfluidics, microarray technology, and nanotechnology facilitate the development of novel detection platforms such as protein and polysaccharide microarrays, lab on a chip, biosensors, and high-throughput sequencing technologies. Such methods will be used for larger applications and will open new avenues in food microbiology. These multiplexed assays will allow the development of tests for the detection of virtually any combination of target sequences in any type of sample that contains nucleic acid material. Genomics, proteomics, and other areas of study concerning microbial cells will provide new opportunities for achieving objectives that were hard to imagine not so long ago. These promising new technologies are making the transition from the research laboratory to routine diagnostic use. Obviously, the new technologies will also need to be carefully evaluated and thoroughly validated for specific applications.

Keywords Protein microarrays • Polysaccharide microarrays • Phenotype microarrays • Biosensors • Nanotechnology • Next-generation high-throughput sequencing technologies

6.1 Other Microarray Technologies

6.1.1 Protein Microarrays

The construction of protein arrays is a significantly greater challenge compared with DNA microarrays due to the lack of a PCR-equivalent amplification process and due to the tridimensional structure of proteins, which requires a wide variety of

G. López-Campos et al., *Microarray Detection and Characterization of Bacterial Foodborne* 115
Pathogens, SpringerBriefs in Food, Health, and Nutrition, DOI 10.1007/978-1-4614-3250-0_6,
© Guillermo Lopez-Campos, Joaquin V. Martinez-Suarez, Mónica Aguado-Urda,
Victoria Lopez Alonso 2012

binding chemistries and specificities. For a review of this technology, see Hall et al. (2007) and Chandra et al. (2011).

Protein microarrays can be classified into three categories based on their possible application to studying foodborne pathogens:

- Functional protein microarrays can detect biochemical activity or highlight antimicrobial mechanisms of action in a pathogen. This type of protein array involves the process of protein expression in the microbial system, followed by their purification and printing onto the array surface by specific immobilization protocols. The process is laborious and time-consuming. Another option is to prepare cell-free expression microarrays, which rely on in situ protein synthesis from their corresponding DNA templates. Many improved cell-free expression microarrays have been developed, such as nucleic acid programmable protein arrays, DNA array to protein array, and HaloTag™, each with its own advantages and limitations (Chandra et al. 2011).
- Analytical protein microarrays detect the presence of specific proteins by exploiting their antibody–antigen properties. The targeted biomolecules are captured on the array surface and detected by specific labeled secondary antibodies, aptamers (short single-stranded oligonucleotides), or affibody molecules (generated by combinatorial protein engineering). Examples of research and diagnostic applications of this type of array are the O-typing of pathogenic *Escherichia coli*, detection of bacterial toxins from a variety of different samples, screening of complex antibody libraries, and epitope mapping (Ehricht et al. 2009).
- Reverse-phase protein arrays are generated by immobilizing the test sample, such as cellular lysate, on the array surface, which is then probed using a detection antibody against the target protein of interest. These antibodies are in turn detected by means of secondary signal amplification. In 2008, Gehring et al. designed an antibody-based microarray for the detection of *E. coli* O157:H7 (Gehring et al. 2008). A sandwich immunoassay format is used through the immobilization of biotinylated capture antibodies on streptavidin-coated glass slides and by using fluorescein-labeled detection antibodies. Protein arrays coupled with bead-based array technology have started to replace ELISA for multiplexed analysis of immunoassays.

The majority of the protein microarray applications to date have employed label-based detection techniques incorporating advances in nanotechnology (radioisotopes, fluorescent dyes, chemiluminescent molecules, quantum dots, gold nanoparticles, nanotubes, and bio-barcodes). The limitations posed by the label-based detection strategies, such as the alteration of surface characteristics of the query molecule, have increased interest in the use of label-free approaches, such as surface plasmon resonance (SPR), atomic force microscopy (AFM), carbon nanotubes, nanowires, micro-electromechanical cantilevers, and mass spectrometry (Ray et al. 2010).

6.1.2 Polysaccharide Microarrays

Microbial polysaccharide microarrays or glycan arrays can be possible after the recent development of new methods for the chemical and enzymatic synthesis of oligosaccharides. The identification and diagnosis of pathogens are achieved because carbohydrates on the cell surface of human cells are used by foodborne bacteria as initial recognition and attachment sites (Liang et al. 2007; Horlacher and Seeberger 2008).

The nondestructive nature of the polysaccharide microarrays allows the pathogen to be harvested and tested for antibacterial susceptibility. These investigations allow microarray-based screening of food samples for pathogen-specific carbohydrate antigens for rapid and specific serological diagnosis (Disney and Seeberger 2004; Blixt et al. 2008).

6.1.3 Phenotype Microarrays

Phenotype microarrays (PMs) are sets of phenotypic assays performed in 100-μl cultures in 96-well microplates containing dried chemicals to assay, for example, C-source, N-source, P-source, and S-source metabolism, biosynthetic pathways, ion effects, osmolarity, pH effects, and chemical sensitivity to ions (Na^+, K^+, Fe^{3+}, Cu^{2+}, chloride, sulfate, chromate, phosphate, vanadate, nitrate, nitrite, selenite, tellurite, etc.). PM technology has been introduced as a tool to characterize the metabolism of pathogens and provide comprehensive cellular profiles that can be used to identify gene function and to validate drug targets and toxicology studies (Fox et al. 2011).

Biolog (Hayward, CA) has developed the Omnilog system to assay nearly 1,200 metabolic and chemical-sensitivity phenotypes in identifying and tracking Gram-negative enteric foodborne pathogens such as *E. coli* O157:H7, *Salmonella*, and *Shigella* (Bochner et al. 2001) based on the detection of bacterial respiration and growth.

6.2 Biosensors

Biosensors are one the most promising solutions for achieving fast, sensitive and real-time detection of foodborne pathogens (Ahmed et al. 2008; Bhunia 2008).

Biosensors are compact analytical devices that combine a specific detection system comprised of enzymes, antibodies, antigens, or nucleic acids with a physical or electrochemical signal transduction system. The interaction of the detecting material with the sample containing the microorganism creates a signal that is proportional to the concentration of the microorganism to be measured. Depending on the method of transduction used, the biosensors are classified as optical, electrochemical, piezoelectric, or magnetic, with the first three currently being the most developed for pathogen detection.

Table 6.1 Some examples of electrochemical biosensors developed to detect foodborne pathogens

Pathogen(s)	Duration of the assay	Sensitivity	Reference(s)
Escherichia coli O157:H7	<1 min	10 CFU/mL	Louie et al. (1998)
Salmonella spp.			
Listeria monocytogenes	1 h	Undetermined	Laschi et al. (2006)
Salmonella spp.	<1 h	Variable	Farabullini et al. (2007)
Listeria monocytogenes			
Escherichia coli O157:H7			
Staphylococcus aureus			
Escherichia coli O157:H7	3–5 h	≤10^3 CFU/mL	LaGier et al. (2007)
Salmonella spp.			
Campylobacter jejuni			
Staphylococcus aureus			
Escherichia coli O157:H7	20 min	1 CFU/mL	Zhao et al. (2007)
Escherichia coli O157:H7	10 min	10 CFU/mL	Maraldo and Mutharasan (2007)
Escherichia coli O157:H7	8 min	61 CFU/mL	Luo et al. (2010)
Escherichia coli O157:H7	30 min	10^2–10^5 CFU/mL	Li et al. (2011)

Electrochemical immunosensors developed for simultaneous multiplexed analysis of pathogens are based on measurements of the electrical change corresponding to interactions of the sample with the surface of the biosensor. These systems are classified according to the observed parameter: current (amperometric), potential (potentiometric), impedance (impedancemetric), or conductivity (conductimetric). Electrochemical biosensors offer several advantages for use in diagnostics, notably by allowing the measurement of interactions in real time and without the need for labeling. Compared to optical biosensors, these biosensors offer the advantage of not being affected by the turbidity of the sample, fluorescence of certain molecules of biological interest, or quenching by certain substances.

Miniaturization and automation of the assays are possible with biosensors. In theory, biosensors can shorten the time between sampling and obtaining results, but their utilization in the future will depend on achieving selectivity and sensitivity comparable to reference methods at a lower cost. Sample volumes in the nanoliter range and the ability to carry out analysis of multiple microorganisms with the same device can reduce the cost of reagents and it can lead to short testing times.

Table 6.1 shows some examples of biosensor prototypes designed to detect foodborne pathogens in a single test in the food industry and other testing situations. A review and future trends in biosensor research activities to detect foodborne pathogens can be found in Arora et al. (2011).

6.3   Nanotechnology for Molecular Diagnostics

Nanomolecular diagnostics refers to the use of nanobiotechnology in molecular diagnostics. Sample preparation is the key parameter that allows nanotechnology to improve the sensitivity and specificity of any analytical method for pathogen detection. Gold nanoparticles have already been introduced to diagnostic testing, and as with quantum dots, their functionalities can be changed with a variety of biomolecules, including antibodies, nucleic acids, peptides, proteins, and carbohydrates. The use of magnetic beads for the purification of DNA or proteins has grown over the last several years and is utilized in a number of commercially available kits. Magnetic nanoparticles made of iron and embedded in copolymer beads can have their polymer coating manipulated to improve nonspecific protein adsorption to the surface of the beads, which increases the specificity of the assay. Nanotechnology also provides label-free detection techniques that can overcome some of the limitations of microarray technologies.

Nanoarrays are the next step in the miniaturization of microarrays. Whereas microarrays are prepared by robotic spotting or lithography, limiting the smallest size to several microns, the introduction of nanoarrays has required further developments in lithography, such as dip-pen nanolithography (DPN). A further in-depth technical discussion is beyond the scope of this chapter. DPN technology has been extended to protein arrays and immunoproteins as well as enzymes (Lynch et al. 2004). BioForce Nanosciences' Protein Nanoarrays contain up to 25 million spots per square centimeter (BioForce 2011).

Lab-on-a-chip systems integrate several processes (from DNA extraction to DNA analysis) within a small, single, portable, fully automated instrument. These devices are developed using microfluidic technology consisting, in many cases, of a fluidic system for sample introduction, a reagent supply, a flow cell, a microarray on a substrate, and a detection system. One of the most relevant implementations of lab-on-a-chip is PCR amplification for multiplexed pathogen detection (Sin et al. 2011). A novel integrated microfluidic capillary electrophoresis lab-on-a-chip has been described by Jung et al. (2011). This method performs rapid, sensitive, and multiplex pathogen detection with "sample-in–answer-out" capability and can be utilized for biosafety testing, environmental screening, and clinical trials. Analyses were performed sequentially within 30 min for multiplex pathogen detection (including *E. coli* O157:H7) at the single-cell level.

6.4   Next-Generation High-Throughput Sequencing Technologies

Next-generation sequencing (NGS) technologies can be utilized to achieve multiplex detection and full identification of pathogens, including subtype determination, characterization of virulence factors, and creation of antimicrobial susceptibility profiles (Fournier-Wirth and Coste 2010). There are several excellent reviews available covering the diverse NGS platforms (MacLean et al. 2009; Su et al. 2011).

The selection of a specific NGS platform depends on the purpose of the application, while striking a balance among sequence read length, running time, and cost:

- The platform 454 GSFLX Titanium (Roche Molecular Systems, CA, USA) is based on pyrosequencing in microreactors on a picotiter plate (Margulies et al. 2005).

This technology generates long sequence reads of about 500 nucleotides in length with run times of hours.

- The Solexa Genome Analyzer (Illumina, Inc., CA, USA) is based on adapter ligation, anchoring to a prepared substrate, in situ PCR amplification, and sequencing using fluorophore-labeled chain terminators (Bennett et al. 2005). This platform generates reads 35–75 nucleotides long, and run times of days.
- The ABI SOLiD (Applied Biosystems, Inc.) uses amplified DNA on beads bound to glass slides. The amplified DNA is sequentially hybridized with short defined oligonucleotides, which contain known nucleotides and a specific fluorophore. Sequencing is done by oligonucleotide annealing and ligation. The SOLiD technology generates reads of 35–50 nucleotides.
- The adoption of single-molecule long-read approaches, such as that offered by Pacific Biosciences, may also reveal more large-scale genomic changes in the near future (Schadt et al. 2010).

When considering how a sequencing technology can be used for specific purposes, it is important to consider three parameters: read length, read quality, and read pairing. If reads are very short, then they are of limited use for the de novo assembly of complete genomes. Although some simple bacterial genome assemblies have been carried out on reads of less than 50 nucleotides, for the vast majority of genomes, assembly is nearly impossible. Nowadays, 454 GSFLX technology is the only one that could be used to do de novo genome sequencing.

Short single reads are still very useful for comparative studies where the aim is to identify single-nucleotide polymorphisms (SNPs) or larger differences between a reference genome and a newly sequenced genome. The low cost of Solexa and SOLiD technologies makes these options excellently suited for SNP analysis.

Overall, all these technologies can be used for the analysis of the transcriptome by next-generation sequencing. Referred to as RNA-seq, this process represents the complete collection of transcribed sequences in a cell. This is usually a combination of coding RNA (mRNA) and noncoding RNA. In the case of bacteria, noncoding RNA ranges from trans-acting small RNAs (sRNA) to cis-acting RNAs (riboswitches), as well as antisense RNAs and protein-interacting RNAs (6S RNA, CsrB-like RNAs). If the genome sequence of the specific strain is not available, it may be possible to utilize a reference sequence from another strain of the same species, although this will result in the loss of some sequence information and an incomplete representation of the genome (Tang et al. 2009; van Vliet and Wren 2009).

High-throughput sequencing has already been applied to the sequencing of whole bacterial genomes (Rogers and Bruce 2010). Metagenomics, or community genomics, is an approach aimed at analyzing the genomic content of microbial communities living in any particular niche, such as the food and the food processing environments. Genomic analysis has been used to circumvent many problems, as it can allow the analysis of nonculturable organisms. Additionally, molecular phylogenetic analysis can be used to study the taxonomic diversity of the organisms present. The added advantage of genomic methods is that the analysis of gene content will also give an indication of the metabolic potential of an environment.

Metagenomic studies have already been applied to several distinct environments such as the human gut (Dominguez-Bello et al. 2011; Zhang et al. 2006) and soil and environmental samples (Mills et al. 2006).

According to Mellmann et al. (2011), NGS technology represents the birth of a new discipline: prospective genomic epidemiology. It is foreseeable that this method will quickly become routine in public health laboratories for surveillance and control of the most severe outbreaks. The rapid publication in the first weeks of the 2011 European outbreak of EHEC O104:H4 of the complete genome sequences of the microorganisms involved is a good example of the application of next-generation sequencing technology for analyzing outbreaks in real time (Bielaszewska et al. 2011; Rohde et al. 2011; Rasko et al. 2011).

Genomics, proteomics, and other areas of study concerning microbial cells will provide new opportunities for achieving objectives that were hard to imagine not so long ago. These promising new technologies are making the transition from the research laboratory to routine diagnostic use.

References

Ahmed N, Dobrindt U, Hacker J, Hasnain SE (2008) Genomic fluidity and pathogenic bacteria: applications in diagnostics, epidemiology and intervention. Nat Rev Microbiol 6:387–394.

Arora P, Sindhu A, Dilbaghi N, Chaudhury A (2011) Biosensors as innovative tools for the detection of food borne pathogens. Biosens Bioelectron 28:1–12.

Bennett ST, Barnes C, Cox A, Davies L, Brown C (2005) Toward the 1,000 dollars human genome. Pharmacogenomics 6:373–382.

Bhunia AK (2008) Biosensors and bio-based methods for the separation and detection of food-borne pathogens. Adv Food Nutr Res 54:1–44.

Bielaszewska M, Mellmann A, Zhang W, Köck R, Fruth A, Bauwens A, Peters G, Karch H (2011) Characterisation of the *Escherichia coli* strain associated with an outbreak of haemolytic uraemic syndrome in Germany, 2011: a microbiological study. Lancet Infect Dis 11:671–676.

BioForce Nanosciences http://www.bioforcenano.com/index.php?id=58. Accessed 1 December 2011.

Blixt O, Hoffmann J, Svenson S, Norberg T (2008) Pathogen specific carbohydrate antigen microarrays: a chip for detection of *Salmonella* O-antigen specific antibodies. Glycoconj J 25:27–36.

Bochner BR, Gadzinski P, Panomitros E (2001) Phenotype microarrays for high-throughput phenotypic testing and assay of gene function. Genome Res 11:1246–1255.

Chandra H, Reddy PJ, Srivastava S (2011) Protein microarrays and novel detection platforms. Expert Rev Proteomics 8:61–79.

Disney MD, Seeberger PH (2004) The use of carbohydrate microarrays to study carbohydrate-cell interactions and to detect pathogens. Chem Biol 11:1701–1707.

Dominguez-Bello MG, Blaser MJ, Ley RE, Knight R (2011) Development of the human gastrointestinal microbiota and insights from high-throughput sequencing. Gastroenterology 140: 1713–1719.

Ehricht R, Adelhelm K, Monecke S, Huelseweh B (2009) Application of protein arraytubes to bacteria, toxin, and biological warfare agent detection. Methods Mol Biol 509:85–105.

Farabullini F, Lucarelli F, Palchetti I, Marrazza G, Mascini M (2007) Disposable electrochemical genosensor for the simultaneous analysis of different bacterial food contaminants. Biosens Bioelectron 22:1544–1549.

Fournier-Wirth C, Coste J (2010) Nanotechnologies for pathogen detection: future alternatives? Biologicals 38:9–13.

Fox EM, Leonard N, Jordan K (2011) Physiological and transcriptional characterization of persistent and nonpersistent *Listeria monocytogenes* isolates. Appl Environ Microbiol 77:6559–6569.

Gehring AG, Albin DM, Reed SA, Tu SI, Brewster JD (2008) An antibody microarray, in multi-well plate format, for multiplex screening of foodborne pathogenic bacteria and biomolecules. Anal Bioanal Chem 391:497–506.

Hall DA, Ptacek J, Snyder M (2007) Protein microarray technology. Mech Ageing Dev 128:161–167.

Horlacher T, Seeberger PH (2008) Carbohydrate arrays as tools for research and diagnostics. Chem Soc Rev 37:1414–1422.

Jung JH, Kim GY, Seo TS (2011) An integrated passive micromixer-magnetic separation-capillary electrophoresis microdevice for rapid and multiplex pathogen detection at the single-cell level. Lab Chip 11:3465–3470.

LaGier MJ, Fell JW, Goddwin KD (2007) Electrochemical detection of harmful algae and other microbial contaminants in coastal waters using hand-held biosensors. Mar Pollut Bull 54: 757–770.

Laschi S, Palchetti I, Marrazza G, Mascini M (2006) Development of disposable low density screen-printed electrode arrays for simultaneous electrochemical measurements of the hybridization reaction. J Electroanal Chem 593: 211–218.

Li D, Feng Y, Zhou L, Ye Z, Wang J, Ying Y, Ruan C, Wang R, Li Y (2011) Label-free capacitive immunosensor based on quartz crystal Au electrode for rapid and sensitive detection of *Escherichia coli* O157:H7. Anal Chim Acta 687:89–96.

Liang PH, Wang SK, Wong CH (2007) Quantitative analysis of carbohydrate-protein interactions using glycan microarrays: determination of surface and solution dissociation constants. J Am Chem Soc 129:11177–11184.

Louie AS, Marenchic IG, Whelan RH (1998) A fieldable modular biosensor for use in detection of foodborne pathogens. Field Anal. Chem Technol 2:371–377.

Luo Y, Nartker S, Miller H, Hochhalter D, Wiederoder M, Wiederoder S, Setterington E, Drzal LT, Alocilja EC (2010) Surface functionalization of electrospun nanofibers for detecting *E. coli* O157:H7 and BVDV cells in a direct-charge transfer biosensor. Biosens Bioelectron 26: 1612–1617.

Lynch M, Mosher C, Huff J, Nettikadan S, Johnson J, Henderson E (2004) Functional protein nanoarrays for biomarker profiling. Proteomics 4(6):1695–1702.

MacLean D, Jones JDG, Studholme DJ (2009) Application of next-generation sequencing technologies to microbial genetics. Nat Rev Microbiol 7:287–296.

Maraldo D, Mutharasan R (2007) 10-minute assay for detecting *Escherichia coli* O157:H7 in ground beef samples using piezoelectric-excited millimeter-size cantilever sensors. J Food Prot 70:1670–1677.

Margulies M, Egholm M, Altman WE et al (2005) Genome sequencing in microfabricated high-density picolitre reactors. Nature 437:376–380.

Mellmann A, Harmsen D, Cummings CA, Zentz EB, Leopold SR, Rico A, Prior K, Szczepanowski R, Ji Y, Zhang W, McLaughlin SF, Henkhaus JK, Leopold B, Bielaszewska M, Prager R, Brzoska PM, Moore RL, Guenther S, Rothberg JM, Karch H (2011) Prospective genomic characterization of the German enterohemorrhagic *Escherichia coli* O104:H4 outbreak by rapid next generation sequencing technology. PLoS One 6:e22751.

Mills DK, Entry JA, Voss JD, Gillevet PM, Mathee K (2006) An assessment of the hypervariable domains of the 16S rRNA genes for their value in determining microbial community diversity: the paradox of traditional ecological indices. FEMS Microbiol Ecol 57:496–503.

Rasko DA, Webster DR, Sahl JW, Bashir A, Boisen N, Scheutz F, Paxinos EE, Sebra R, Chin CS, Iliopoulos D, Klammer A, Peluso P, Lee L, Kislyuk AO, Bullard J, Kasarskis A, Wang S, Eid J, Rank D, Redman JC, Steyert SR, Frimodt-Moller J, Struve C, Petersen AM, Krogfelt KA, Nataro JP, Schadt EE, Waldor MK (2011) Origins of the *E. coli* strain causing an outbreak of hemolytic-uremic syndrome in Germany. N Engl J Med 365:709–717.

Ray S, Chandra H, Srivastava S (2010) Nanotechniques in proteomics: current status promises and challenges. Biosens Bioelectron 25:2389–2401.

Rogers GB, Bruce KD (2010) Next-generation sequencing in the analysis of human microbiota: essential considerations for clinical application. Mol Diagn Ther 14:343–350.

Rohde H, Qin J, Cui Y, Li D, Loman NJ, Hentschke M, Chen W, Pu F, Peng Y, Li J, Xi F, Li S, Li Y, Zhang Z, Yang X, Zhao M, Wang P, Guan Y, Cen Z, Zhao X, Christner M, Kobbe R, Loos S, Oh J, Yang L, Danchin A, Gao GF, Song Y, Li Y, Yang H, Wang J, Xu J, Pallen MJ, Wang J, Aepfelbacher M, Yang R (2011) Open-source genomic analysis of Shiga-toxin-producing *E. coli* O104:H4. N Engl J Med 365:718–724.

Schadt EE, Turner S, Kasarskis A (2010) A window into third-generation sequencing. Hum Mol Genet 19:227–240.

Sin ML, Gao J, Liao JC, Wong PK (2011) System integration – a major Step toward Lab on a Chip. J Biol Eng 5:6–28.

Su Z, Ning B, Fang H, Perkins R, Tong W, Shi L (2011) Next-generation sequencing and its applications in molecular diagnostics. Expert Rev Mol Diagn 11:333–343.

Tang F, Barbacioru C, Wang Y, Nordman E, Lee C, Xu N, Wang X, Bodeau J, Tuch BB, Siddiqui A, Lao K, Surani MA (2009) mRNA-Seq whole transcriptome analysis of a single cell. Nat Methods 6: 377–382.

van Vliet AH, Wren BW (2009) New levels of sophistication in the transcriptional landscape of bacteria. Genome Biol 10:233.

Zhang W, Qi W, Albert TJ, Motiwala AS, Alland D, Hyytia-Trees EK, Ribot EM, Fields PI, Whittam TS, Swaminathan B (2006) Probing genomic diversity and evolution of *Escherichia coli* O157 by single nucleotide polymorphisms. Genome Res 16:757–767.

Zhao W, Wang L, Tan W (2007) Fluorescent nanoparticle for bacteria and DNA detection. Adv Exp Med Biol 620:129–135.

Index

G. López-Campos et al., *Microarray Detection and Characterization of Bacterial Foodborne*   125
Pathogens, SpringerBriefs in Food, Health, and Nutrition, DOI 10.1007/978-1-4614-3250-0,
© Guillermo Lopez-Campos, Joaquin V. Martinez-Suarez, Mónica Aguado-Urda,
Victoria Lopez Alonso 2012